3D Bioprinting in Medicine

Murat Guvendiren
Editor

3D Bioprinting in Medicine

Technologies, Bioinks, and Applications

Springer

Editor
Murat Guvendiren
Otto H. York Department of Chemical and Materials Engineering
New Jersey Institute of Technology
Newark, NJ, USA

ISBN 978-3-030-23905-3   ISBN 978-3-030-23906-0 (eBook)
https://doi.org/10.1007/978-3-030-23906-0

This Springer imprint is published by the registered company Springer Nature Switzerland AG
The registered company address is: Gewerbestrasse 11, 6330 Cham, Switzerland

Preface

This book is intended to provide an introduction to bioprinting, including bioprinting technologies, bioinks, and applications of bioprinting. This book is composed of seven chapters contributed by experts in their fields.

Chapter 1 focuses on the in-depth understanding of the bioprinting technologies, including extrusion-based bioprinting, droplet-based bioprinting, and bioprinting achieved through directed application of energy. A detailed discussion on the strengths and weaknesses of these techniques is given. This chapter also summarizes advances in bioprinting to address challenges in bioprinting technologies.

Chapter 2 outlines the major concepts in the design and formulation of bioinks, including key parameters important in material selection to expand the range of bioink and, hence, bioprinted construct properties for desired applications.

Chapter 3 gives an overall summary of potential clinical applications of bioprinting, followed by Chaps. 4–6, each focusing on a specific application. Chapter 4 discusses bioprinting of vascularized constructs and their utilization as scaffolds in tissue regeneration and as platforms for drug discovery and testing. In Chap. 5, the focus is on the applications of bioprinting in the clinical cardiovascular medicine, including surgical models, cardiac patches, computational and theoretical models, heart valves, and stents as well as in vitro tissue models. Chapter 6 summarizes bioprinting of in vitro bone models for understanding cancer cell behavior and microenvironment.

Chapter 7, the final chapter, aims to address the ethical, safety, and regulatory aspects of bioprinting, specifically related to bioprinting technology, bioinks, and bioprinted constructs.

Newark, NJ, USA Murat Guvendiren

Preface

Acknowledgments

Dr. Guvendiren (the editor) would like to thank the New Jersey Institute of Technology for the support during the preparation of this book, acknowledges the support from the National Science Foundation (NSF, USA) Division of Materials Research (DMR) Award Number 1714882, and is grateful to Shen Ji and Andrew House for their assistance during the review process.

Contents

Contributors

Nureddin Ashammakhi Center for Minimally Invasive Therapeutics (C-MIT), University of California Los Angeles, Los Angeles, CA, USA

California NanoSystems Institute (CNSI), University of California Los Angeles, Los Angeles, CA, USA

Department of Radiological Sciences, David Geffen School of Medicine, University of California Los Angeles, Los Angeles, CA, USA

Biotechnology Research Center, Libyan Authority for Research, Science and Technology, Tripoli, Libya

Hafiz Busari Department of Materials Science and Engineering, Lehigh University, Bethlehem, PA, USA

Paula Camacho Department of Bioengineering, Lehigh University, Bethlehem, PA, USA

Alexander Cetnar Department of Biomedical Engineering, Emory University and Georgia Institute of Technology, Atlanta, GA, USA

Lesley W. Chow Department of Bioengineering, Lehigh University, Bethlehem, PA, USA

Department of Materials Science and Engineering, Lehigh University, Bethlehem, PA, USA

Sanskrita Das Department of Creative IT Engineering, Pohang University of Science and Technology, Pohang, Republic of Korea

Anusha Elumalai Molecular Science and Nanotechnology Program, Louisiana Tech University, Ruston, LA, USA

School of Biological Sciences, Louisiana Tech University, Ruston, LA, USA

Christopher B. Highley Department of Biomedical Engineering, University of Virginia, Charlottesville, VA, USA

Ahmed Humayun Molecular Science and Nanotechnology Program, Louisiana Tech University, Ruston, LA, USA

Jinah Jang Department of Creative IT Engineering, Pohang University of Science and Technology, Pohang, Republic of Korea

Department of Mechanical Engineering, Pohang University of Science and Technology, Pohang, Republic of Korea

School of Interdisciplinary Bioscience and Bioengineering, Pohang University of Science and Technology, Pohang, Republic of Korea

Bryanna Lima Department of Biomedical Engineering, Emory University and Georgia Institute of Technology, Atlanta, GA, USA

Yangyang Luo Molecular Science and Nanotechnology Program, Louisiana Tech University, Ruston, LA, USA

David K. Mills School of Biological Sciences, Louisiana Tech University, Ruston, LA, USA

Center for Biomedical Engineering and Rehabilitation Science, Louisiana Tech University, Ruston, LA, USA

Ippokratis Pountos Academic Department of Trauma and Orthopaedics, Leeds Teaching Hospitals, University of Leeds, Leeds, UK

Chapel Allerton Hospital, Leeds Teaching Hospitals, Leeds, UK

Kelly B. Seims Department of Materials Science and Engineering, Lehigh University, Bethlehem, PA, USA

Vahid Serpooshan Department of Biomedical Engineering, Emory University and Georgia Institute of Technology, Atlanta, GA, USA

Department of Pediatrics, Emory University School of Medicine, Atlanta, GA, USA

Children's Healthcare of Atlanta, Atlanta, GA, USA

Nazzar Tellisi Academic Department of Trauma and Orthopaedics, Leeds Teaching Hospitals, University of Leeds, Leeds, UK

Chapel Allerton Hospital, Leeds Teaching Hospitals, Leeds, UK

Andrea Theus Department of Biomedical Engineering, Emory University and Georgia Institute of Technology, Atlanta, GA, USA

Department of Pediatrics, Emory University School of Medicine, Atlanta, GA, USA

John W. Tolbert Department of Polymer Science and Engineering, Lehigh University, Bethlehem, PA, USA

Martin Tomov Department of Biomedical Engineering, Emory University and Georgia Institute of Technology, Atlanta, GA, USA

Department of Pediatrics, Emory University School of Medicine, Atlanta, GA, USA

Agastya Vaidya Department of Biology, Emory University, Atlanta, GA, USA

Emory University and Georgia Institute of Technology, Atlanta, GA, USA

About the Editor

Murat Guvendiren is the principal investigator of the Instructive Biomaterials and Additive Manufacturing Laboratory (IBAM-Lab) and an assistant professor in the Chemical and Materials Engineering Department at the New Jersey Institute of Technology (NJIT). He earned his B.S. and M.S. degrees in Metallurgical and Materials Engineering from the Middle East Technical University in Ankara, Turkey. He earned his Ph.D. degree from the Department of Materials Science and Engineering at Northwestern University. He did a postdoctoral training at the University of Pennsylvania. Before joining to NJIT, he was a research assistant professor at the New Jersey Center for Biomaterials, Rutgers University. His research focuses on the development of novel bioink formulations and bioprinting of in vitro tissue/disease models.

About the Editor

Chapter 1
3D Bioprinting Technologies

Christopher B. Highley

Abstract Bioprinting technology offers capabilities for the design and fabrication of biological and tissue structures. Some of the field's products are already impacting human health. Research to increase the complexity and functionality of bioprinted structures through innovation in bioprinting hardware, techniques, and materials continues to expand, with the aim of ultimately producing patient-specific tissue constructs. This chapter offers an introduction to the field, first by considering the emergence of bioprinting in relationship to enabling technologies. Because of the intense interest and continuing growth of the bioprinting, specific terminology and potential points of confusion across academic publications are discussed. The three main classes, or modalities, of bioprinting technologies are covered—extrusion-based bioprinting, droplet-based bioprinting, and bioprinting achieved through directed application of energy, such as light, to a bioink. The strengths and weaknesses of these techniques are discussed next, to highlight the potential of varying approaches and to point to opportunities for advances. Finally, some recent advances in bioprinting that are addressing challenges that exist in the three modalities are introduced to highlight innovative approaches that have and will continue to advance the field beyond limitations in the processes used.

Keywords 3D bioprinting · Bioprinting technology · Bioink · Historical perspective

1.1 Introduction

A motivating vision for work within the field of bioprinting is realizing medical technology that seems fantastic, and describing it sounds like excerpting an idea from a work of science fiction. In this vision, bioprinting would yield a machine

The original version of this chapter was revised. The correction to this chapter is available at https://doi.org/10.1007/978-3-030-23906-0_8

C. B. Highley (✉)
Department of Biomedical Engineering, University of Virginia, Charlottesville, VA, USA
e-mail: ch2qm@virginia.edu

M. Guvendiren (ed.), *3D Bioprinting in Medicine*,
https://doi.org/10.1007/978-3-030-23906-0_1

with the capability to create patient-specific, living tissue on demand. The patient might be someone who has experienced organ failure, a soldier or a civilian in a war zone who has suffered a traumatic loss of large volumes of tissue, or an individual undergoing treatment for a disease that required a significant resection surgery. At the site of treatment, a patient-specific 3D computer model of the lost tissue would be created from images of the injury, and this 3D model would be transmitted to 3D bioprinting hardware, which would then—as if out of nothingness—create a living organ or tissue replacement. From the machine would come a fully functional biological organ or tissue system, which would be compatible with, and a perfect fit for, the patient in need (Fig. 1.1).

Presently, this remains a vision. But given adequate control over how cells and biomaterials are positioned in 3D space, and sufficient knowledge of what structures need to be established to yield a functional volume of cells and materials acting in concert to perform the physiological functions of tissue—given these things, it stands to reason that we would be able to fabricate a living system that can augment or replace lost tissue or even build living tissue itself. Bioprinting is one technology that allows us to position cells and materials with high resolution in 3D space, an important prerequisite for achieving this goal. Complementary efforts in basic science and in engineering and applied science, often using biofabrication technologies, will yield advances in technology and fundamental science that occur simultaneously, and often synergistically. For example, by allowing us to create systems where we can ask how cells respond to given features of their microenvironments, new technologies enable us to address questions within the realms of medicine and fundamental science [2–4]. In turn, this knowledge will inform the development of bioprinting and related technologies, allowing us to understand what materials and cells we want to print and to specify technical parameters, such as at what resolution a bioprinting technology must be able to print cells and materi-

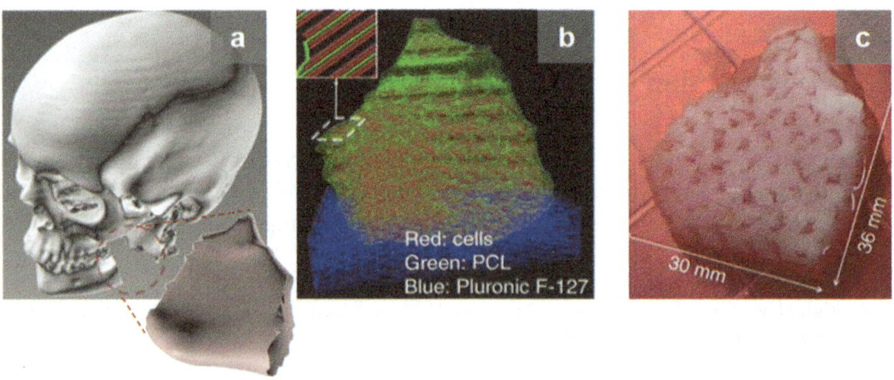

Fig. 1.1 (**a**) A patient-specific injury can be scanned and turned into a 3D computer model (see zoomed object that would fill the bony defect in the mandible), which (**b**) in turn is used to generate print commands (here, three different colored print paths—see zoomed inset—indicate where three different inks will be printed), (**c**) resulting in a bioprinted replacement tissue (adapted from Kang et al. Nat Biotech 2016 [1])

als throughout a given tissue construct to yield appropriate structure and function in the final product.

Some of the biggest questions facing bioprinting do not belong to bioprinting alone. Understanding the many factors that influence cellular organization into functional tissue and how this dictates control of structure and positioning of cells and materials within tissue constructs represent big questions that are central to efforts in broader fields of tissue engineering and regenerative medicine. The attraction of, and excitement surrounding, bioprinting as means of fabricating tissue engineering structures is due in part to its capabilities and potential as a technology for addressing these big questions. Even its current capabilities enable the fabrication of tissue engineering constructs with complexity that allows us to probe fundamental biological and physiological questions, while continuing advances are bringing us closer to solving grand biomedical challenges.

1.2 Contributing Technologies and a Historical Perspective on 3D Printing, Tissue Engineering, and Bioprinting

Bioprinting has emerged at the intersection of larger fields: additive manufacturing, tissue engineering + regenerative medicine, and biofabrication (Fig. 1.2a). In particular, additive manufacturing (AM) and tissue engineering (TE) technologies

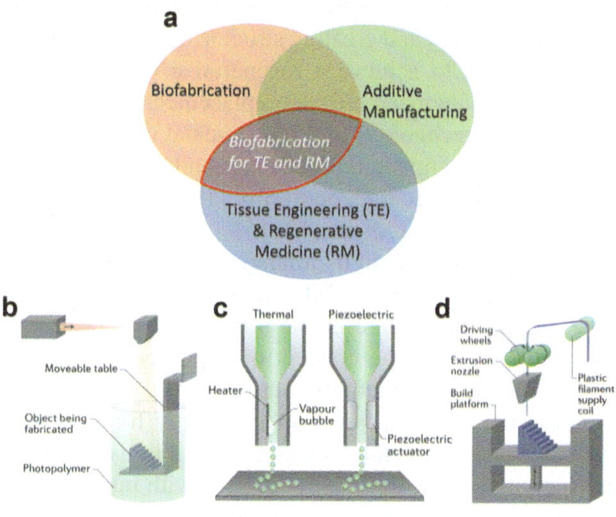

Fig. 1.2 (**a**) A Venn diagram illustrating the relationships between the fields of TE, AM, and biofabrication (circle sizes are arbitrary and not representative of any attribute of the fields). (**b**) Schematic illustration of general features of an SLA process. (**c**) Schematic illustration of general features of inkjet printing. (**d**) Schematic illustration of general features of FDM fabrication((**a**) adapted from Groll et al. Biofabrication 2016 [5], (**b–d**) adapted from Moroni et al. Nat Rev Mater 2018 [6])

together largely define bioprinting, and both emerged in the late 1980s and through the 1990s, with work at the interface of AM and TE occurring almost from the outset. While both AM and TE have roots in work predating the late 1980s [7, 8], the first patent for a AM technology—stereolithography—where ultraviolet light was used to cure one layer of a polymer on top of another using a computer-controlled light beam and hardware, was issued in 1986. At roughly the same time, the laser-based sintering of select regions of a layer of powder, followed by the repeated application of successive layers of powder and sintering to establish a 3D structure was developed, including using metal powders. Fused deposition modeling (FDM), where heat was used to extrude filaments of thermoplastic or metal, layer upon layer, to print a 3D structure was also patented at the turn of the decade [8]. Inkjet printing had been established for decades at this point, and so by the end of the 1980s, the foundations were laid for extrusion-, energy- (light-), and droplet-based printing techniques commonly used in bioprinting today (Fig. 1.2b–d).

Efforts to apply engineering to create living tissues are viewed as having begun in earnest in the late 1980s [7] and reached broader awareness in the early 1990s, with a paradigm for engineering tissues being established. This paradigm, which remains central to TE, and also bioprinting efforts, describes the use of biomaterial scaffolds designed to support relevant cell types toward the regeneration of a deficient tissue in a patient [9]. Cells might be placed on (or within, in the case of hydrogels) the scaffold structure in vitro, and the scaffold might include biochemical signals or other cues to direct cells associating with it toward regenerative phenotypes. Ultimately, the cell-material construct would be implanted into the patient which would develop into, or guide the development of, a new tissue. The challenges surrounding identifying and creating structures with the correct cells, materials, and signals necessary to achieve a regenerative outcome remain central to research. Fully functional engineered tissue replacements are still lacking for most tissue types; however, TE technology and fundamental research understanding cellular behaviors have had synergistic relationships, and technologies associated with TE are important in drug screening platforms [10, 11], stem cell and organoid research [12], and commercial products [13–15].

Though tissue engineering was in its infancy as a field in the late 1980s and the use of printing technologies to create 3D structures of cells within hydrogels was also more than a decade away, there is one remarkable paper describing not only two bioprinting modalities, but the concept of "micropositioning cells" to create "two- and three-dimensional synthetic tissues" [16, 17]. Before tissue engineering or biofabrication existed as fields, this paper not only foresaw but demonstrated advanced tissue engineering technologies that are still in use today. Capabilities to position cells at high resolution were demonstrated with inkjet printing of a bioink and through the use of stereolithographic modification of a substrate. A method of creating multiple layers of cells was all described, although it would be many years before any of these techniques would see broad use (Fig. 1.3).

The 1990s continued to see the development of AM and TE technologies both independently of each other, though with preliminary steps in bioprinting. In AM, application and refinement of newly established technology occurred alongside the

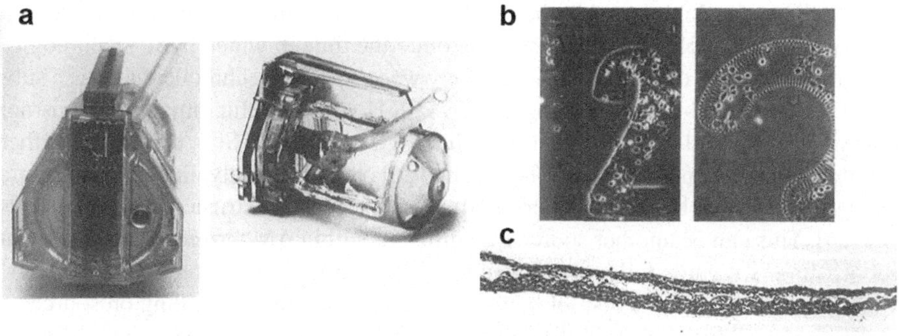

Fig. 1.3 (**a**) An inkjet printhead adapted for the printing of biofunctional molecules (e.g., fibronectin) from the front (left) and side (right) where a hole cut into the side to remove the original ink bladder and replace it with a silicone tube into which the fibronectin solution was fed. (**b**) Photolithographic patterning of cell adhesive regions. (**c**) Construction of a 3D tissue formed by layers of cells, patterned on thin collagen sheets using inkjet methods, and then "glued together" with collagen. This work was done in 1988 (adapted from Klebe Exp Cell Res 1988 [16])

continued expansion of AM methods. Robocasting [18, 19] was developed, and it has since been distinguished with respect to other extrusion-based techniques, in that it has employed materials that would undergo flow and stabilization simply through the application and cessation of flow-driving forces. Two-photon, femtosecond laser-based technology was also identified as a microfabrication technology [20]—a method that would enable additive manufacturing approaching the smallest of scales, including below the diffraction limit of light [21].

It should be noted that usage of the term AM throughout this chapter is chosen over 3D printing (3DP), despite the modern usage of 3DP, which often equates the two. This choice is made because in the historic origins of AM technology, *3D printing* referred specifically to technology that deposited binders onto sequential layers of powders, and AM and *solid freeform fabrication* (SFF) and *rapid prototyping* were more general terms that encompassed a broad cross section of technologies [22, 23]. SFF, for example, is also a general term and encompasses all techniques that deposit material and/or direct energy to material to be cured or otherwise bound together to manufacture an object [24]. According to the terminology that defined 3D printing as the deposition of binders onto powders—and that was contemporaneous to fused deposition modeling's emergence—FDM was an SFF technique, but FDM was not 3D printing [22]. In modern usage, 3DP practically equates to the historic term, SFF, and thus FDM is considered a type of 3DP. This can be confusing, and we will aim to use precise terminology in this chapter. However, it is unlikely that the common usage of "3D printing" to refer to a broad cross section of AM technologies will disappear, especially given that the hardware that creates a 3D object is often referred to as a "printer," and the term "bioprinting" (discussed more in the next section) encompasses a broad range of technologies to "bioprint" 3D objects that extend well beyond the original "3D printing."

It bears mentioning here that descriptions of *direct write* or *direct ink writing* (DIW) technologies historically carry meaning very similar [25–27] to SFF. Direct write/DIW technologies enable the fabrication of structures without masks [26] (in contrast to lithography approaches) and around the time in which these technologies were emerging, the original 3DP technology was a specific technique that was a subset of DIW- or SFF-type technologies [23, 25]. However, in literature, DIW has not always been viewed as a broad set of technologies similar to SFF, but instead either equated with a technique called robocasting [28] or, alternatively, listed as a separate technology from robocasting where both are subtypes of extrusion-based methods [22, 24]. This can be another source of confusion within AM terminology, given that robocasting is regarded elsewhere as one subtype among many DIW technologies. Clearly, terminology evolves, for better or worse, and varies depending on source.

Focuses of TE research in the 1990s included scaffold processing and different tissue targets [29]. Toward bioprinting, initial work was undertaken by pioneers that bridged the still young fields of tissue engineering and SFF. For example, in some of the earliest work on what would now be considered bioprinting, the 3DP-based ("3DP" in its original meaning) deposition of a chloroform binder onto sequential poly(lactic acid) powder layers yielded scaffolds for drug release [30] or bone regeneration [31]. In forward-looking work, this technology was also used to spatially control regions where cells might adhere within a scaffold, toward controlling the arrangements of multiple cell types within microarchitectured scaffolds [32]. Selective laser sintering of hydroxyapatite and calcium phosphate particles coated with a polymeric binder yielded a potential bioceramic for orthopedic implantation [33]. Stereolithography was used to fabricate a scaffold from a slurry consisting of hydroapatite particles, low viscosity acrylate, and a photoinitiator. FDM [34] deposition of poly(caprolactone), a common material in TE scaffolds, to create a porous construct which could be seeded with cells for TE, followed soon after. Innovative, non-SFF approaches to TE at the microscale were pioneered during this time, and microfabrication began to come into its own [35], including, for example, influential demonstrations of multiple cell types arranged by lithography-based techniques [36] and of the critical importance of features of cellular microenvironments [37]. Consideration of how technology would enable microscale engineering would remain prominent in the development biofabrication and bioprinting technology to come. The printing of cells using a laser-guided process, similar to optical/laser tweezers, was demonstrated in 1999 [38].

The early 2000s saw the continued development of AM capabilities, often through innovation on the material (as opposed to hardware) side of AM technology that expanded possibilities for AM and bioprinting. For example, new capabilities were demonstrated enabling the printing soft hydrogel materials [39] and the use of fugitive inks in AM to print complex channel structures [40]. In the mid-2000s, the patent on FDM expired [8], which would be important for the spread of AM and positively influence activity in bioprinting as an area of research in recent years. After the patent expiration, popular (non-specialist) open source AM hardware movements were initiated—in general discussion, these technologies were often referred to as open source 3DP, contributing to the blurring of the meaning of the term 3D printing. The term *open source* referenced a software (non-AM) movement that emerged among computer programmers and software developers in the 1990s.

Proponents of open source software aimed to democratize computer technology, particularly through *open*—i.e., collaborative and public—sharing and development of *source* code for software. That ethos was shared by those initiating and participating in open source AM movements, where a central goal was disseminating this powerful manufacturing technology, which would allow digital sharing and subsequent fabrication of physical products and tools with anyone, anywhere on the planet.

The open source RepRap and Fab@Home projects, which originated at the University of Bath and Cornell University [41, 42], respectively, were successful in spreading affordable kits and catalyzing the formation of communities of "makers" who continue to engage with and expand these open source technologies online. Many now working with AM technology (often referred to as "3D printers" or "3D printing" in popular, but historically confusing, usage) and bioprinters, in academic as well as commercial and entrepreneurial settings, first experienced AM through do-it-yourself kits or participate in Makerspaces that bring together people within communities who are interested in these technologies. The accessibility of the hardware and communities, both online and in person, has helped enable researchers whose experience might be more heavily based in tissue engineering or biomaterial research to begin experimenting and innovating with AM hardware [43].

During this same time, research efforts in tissue engineering were expanding understanding and possibilities for engineering extracellular factors that influence cell fate. Work focused on controlling extracellular complexity to influence cell phenotypes within cell-material constructs increased, as did research in related areas, such as at the interface of TE and stem cell biology [44]. In an example of the synergistic relationship between fundamental and applied science, fundamental insights into how mechanical factors influence stem cell fates [45] led to rapid expansion of research in mechanobiology and consideration of mechanical factors in tissue engineered constructs [46]. Dynamic materials for TE and regenerative medicine became more sophisticated [47], and technologies converged, for example, in enabling stem cell-based constructs where temporal control over material properties modified cellular phenotypes [48].

At the interface of fabrication and TE, new techniques were invented to assemble tissue constructs, often toward the goal of using technology to control cellular microenvironments in cell-material constructs [6, 49, 50]. Techniques included those to position cells within hydrogels [51] and to assemble individual hydrogel [52] or cellular [53] components into larger constructs [54] in bioassembly [5] approaches. During this decade, increasing interest and excitement surrounding bioprinting technology was thus related to broad interest in biofabrication approaches to address challenges within TE [29].

Considering bioprinting- and SFF-type technologies applied to TE specifically, trailblazing work in early 2000s would lead establishing bioprinting as an area of research within the broader field of biofabrication. An expansion of techniques, including SFF technology [55] that had thus far seen limited or no use in TE or soft material applications, such as robotic dispensation [56, 57], were used in the creation of biofunctional or bioactive scaffolds for TE. Efforts also began to focus around the concept of bioprinting tissues and organs. In place of polymer-rich materials as inks, in the early 2000s, cell aggregates and spheroids were used as bioinks, deposited initially using a modified inkjet printer fitted with luer-lock needles [39, 58] in a hybrid droplet-based technology that pointed toward the eventual use of extrusion-based technologies in spheroid printing. The printing of cell spheroids was introduced as an approach to tissue and "organ printing" [58, 59], where high cell numbers would be required to achieve cell densities seen in vivo. This cell aggregate-based approach to formulating and printing inks eventually led to the founding of the company Organovo. The idea that cells might be printed was explored using an aerosol-based technique, with the idea that it might be used to seed cells into a 3D scaffold for 3D tissue [60]. A laser-based droplet formation technique (laser-induced forward transfer) was shown to be able to print cells on a 2D surface [61, 62].

Other biofabrication technologies were also introduced in the early 2000s with biomedical (e.g. TE) applications in mind at their inception, reflecting biomedical need motivating adaptations of SFF processes. For example, new capabilities for extruding hydrogel materials were introduced in a specialized form of extrusion called bioplotting [57]. A BioAssembly Tool, similar to what might now be considered a custom built extrusion-based printer, demonstrated both pneumatic and plunger-driven extrusion of multiple cell-containing bioinks [63, 64]. Other extrusion-based methods for AM of tissue scaffolds were explored as well, both with and without cells [65], including early uses of alginate-based bioinks [66, 67]. Lithographic approaches progressed from patterning multiple cell types without a printer [68] and printing hydrogels without cells [69] to first attempts using SLA to fabricate cell-containing scaffolds [70]. During this time, the strengths of two-photon laser-based lithography were harnessed to modify cellular microenvironments to control cell behaviors [71–74].

While advances in material development for bioprinting will be covered in more detail in the next chapter, it is worth noting that soft, hydrogel materials, which are important TE [75], are also particularly important in the bioprinting of cell-containing inks. Progress in the development of new and existing methods for the extrusion and droplet-based printing of these materials was, and remains, an important focus of bioprinting research. In the 2000s, inkjet printing was demonstrated in printing hydrogels and cell-containing bioinks [76–79], and reports on the extrusion of hydrogel-based bioinks followed from the expanding field. However, translating many hydrogel systems developed for TE to bioprinting remained challenging owing to printing requirements of flow and rapid stabilization, and this remains an enduring challenge in bioprinting [80, 81]. Hydrogels that could undergo rapid cross-linking in the presence of a chemical or thermal cue—in particular alginate [66, 67] and gelatin [82], respectively—became well-established as important bioinks in bioprinting research.

In the beginning of the 2010s, bioprinting was well-established, and the field grew [83] alongside other applications of AM in science and engineering. Each bioprinting technology (e.g., extrusion, droplet-based, and light-based, to name some major categories) has inherent strengths and limitations, and researchers have worked to address challenges toward realizing bioprinting's potential, as will be discussed in sections specific to the technologies below. Progress continues to be made in applications of bioprinting technology to address major challenges in TE and regenerative medicine—achieving functional tissue structures, or using the technologies for printing vasculature, for example. These areas of application of the technology will also be discussed in greater depth in later chapters, and they represent significant research efforts undertaken as bioprinting technologies have advanced. In terms of the emergence of new bioprinting modalities, toward addressing the challenge of attaining increasing control over resolution and microscale, the emergence in recent years of melt electrowriting [84, 85] is noteworthy. Melt electrowriting offers unique strengths through combining strengths of AM with electrospinning, enabling deposition at microscale resolution of filaments with diameters possible in the 10s of microns to single micron range.

Looking forward, the continued, synergistic development of hardware and material technology has been seen in recent years. Adaptations will be made to hardware and the printing process to facilitate the printing of bioinks that do not have ideal print properties, and vice versa—materials will be modified to render them printable within using existing AM technologies. Within existing bioprinting technology, we now have multiscale capabilities—high-resolution capabilities in addressing 3D space and the ability to address that space across macro-length scales—that even the least expensive printers can achieve, but which remain as yet not fully utilized. Identifying materials, or formulations of materials, with properties useful for printing will continue to represent an important direction for ongoing research. Combining bioprinting modalities will similarly offer opportunities to design processes that bring multiple tools to bear on complex problems, where a single technology might not be optimal for all facets of a problem. The strengths of SFF/AM approaches in application to TE challenges have not become any less attractive since the technologies' invention, and—as will be discussed in this book—these strengths will continue to address challenges within TE [29, 86] in order to advance the field's capabilities to design and build functional tissue structures.

1.3 What Defines Bioprinting Among Related Technologies?

With time, technologies capable of positioning cells and materials in 3D space toward creating tissue constructs of defined compositions have become increasingly sophisticated and diverse. Thus, defining *bioprinting* among related technologies is important not only in the context of this chapter but more broadly within research communities, especially in terms of communicating results and ensuring that the meanings behind terminology used can be broadly understood [5, 87, 88]. Moreover, with respect to the history of bioprinting, the potential confusion surrounding the terms 3D printing, solid freeform fabrication, direct ink writing, robocasting, and extrusion-based printing was discussed. Difficulties with understanding distinctions that are drawn through word choice adds to challenges in understanding the field's work, especially for scientists and engineers new to AM and bioprinting technology.

According to current proposed definitions, bioprinting is a subset of biofabrication technology, where *biofabrication* is defined as "the automated generation of biologically functional products with structural organization from living cells, bioactive molecules, biomaterials, cell aggregates such as micro-tissues, or hybrid cell-material constructs, through Bioprinting or Bioassembly and subsequent tissue maturation processes" [5]. *Bioprinting* specifically refers to fabrication using "computer-aided transfer processes for patterning and assembling living and non-living materials with a prescribed 2D or 3D organization in order to produce bioengineered structures serving in regenerative medicine, pharmacokinetic and basic cell biology studies" [5, 89]. *Bioassembly* refers therefore to non-bioprinting techniques used to assemble a larger cell-material construct from smaller components [5, 55, 90, 91].

According to this terminology, fabrication by bioprinting includes using AM technology to print cells, biomaterials, and bioactive molecules, but it does not require that printed material include cells. An acellular construct could therefore be the product of bioprinting if the intended use is biological, where, for example, cells that are seeded onto—or infiltrate—the construct would be directed by the printed scaffold to organize into a biological structure or achieve a biological outcome. It is worth noting that bioprinted materials can therefore include both *biomaterial inks* and *bioinks*. Biomaterial inks are quite simply biomaterials that are designed to be processed using AM technologies. Biomaterial inks might include soft hydrogel materials deposited using extrusion or droplet-based printing [92]—or bioactive hydroxyapatite-based scaffolds fabricated from powders using printed binders and sintering [93]. Bioinks refer specifically to a cell-containing material deposited through an AM process [77, 87, 94, 95]. In fact, a bioink may be composed almost exclusively of cells with no other material component—historically the term bioink emerged in conjunction with technology for printing cell aggregates [96]. Thus, while bioprinting does not always employ a bioink, an AM process using a bioink will be bioprinting.

Defining *bio*printing as AM processes that can use biomaterial inks that do not contain cells (as long as the final application is biological), but *bio*inks as printed materials that must include cells, is a potential point of confusion. These terms are grounded in the historic evolution of the field. Research in using of AM technologies toward biological applications—for example, in the creation of early, acellular TE scaffolds—had been ongoing for a number of years by the early 2000s when both the terms bioprinting and bioinks came into use. Notably, neither "bioink" nor "bioprinting" was used by some of the field's pioneers in a 2003 publication describing "organ printing" [97], but in 2004 both terms were in use as these early researchers, and others, met in a first international workshop in forward-looking efforts to "consolidate this new direction in bioengineering" and to build scientific community around it [76]. The term bioink, which emerged around this time, was deliberately designed to indicate a printable ink that consisted of a biomaterial and cells or that contained only cells—differentiating this innovation from acellular approaches in use at that point. Again, owing to potential confusion surrounding the prefix "bio-," there may exist a need for further definition and standardization of terminology. But more recent acellular printing examples that would be considered bioprinting include work on the droplet-based printing of a hydroxyapatite scaffold directly into a cranial defect to apply additive manufacturing technology directly in a medical application [98].

This chapter focuses on bioprinting technology for the printing of cell-containing structures and bioinks, which should comply with all usage of the term *bioprinting* elsewhere and remain relevant should an official standard emerge. Because of the potential for confusion, and given the lack of an industry standard definition, such as established by ASTM, it remains important to be careful to pay attention to meanings and methods when reading literature in the field of bioprinting. It is important to be aware of efforts to define a standard nomenclature [5, 87, 99], and this chapter aims to adhere to those definitions but tries to be clear in stating when a particular bioprinting approach is acellular. Should standard terminology eventually be defined, the reader is advised that certain techniques now considered "bioprinting"—here and elsewhere—may fit under an alternate category of biofabrication in the future.

1.4 Bioprinting Workflow

In AM, a physical object is created through a process that translates a digital representation of a physical object into a tangible product. In bioprinting, the process is generally the same, although there might be technical considerations surrounding the use of soft materials or bioinks that would have to be considered [55]. In a typical AM process, a 3D computer representation or model of the final product is created using computer-aided design (CAD) software or generated from imaging data. Next—in the case of layer-by-layer approaches, which represent the majority of AM, printing-type, processes—this model would be computationally sectioned into layers (or *sliced*). Based on the structure of each layer, a series of commands would be generated that describe how the printer should move, and how material should be deposited during these movements. These commands are then transferred to a printer, which would then process them to actuate movements of printheads, build surfaces, and other hardware components, thereby depositing layer upon layer of material(s), according to the print commands, until the desired structure is achieved (Fig. 1.4a).

As described above, when a product is being manufactured from a computer design, that object must first be constructed in silico as 3D model. CAD software might be used to first model a part, although, in bioprinting, the goal is often a substitute or replacement for native biology. In the ultimate application of bioprinting, the part would be a patient-specific tissue construct, as has been demonstrated in the printing of hydroxyapatite scaffolds for bone regeneration, which have been designed to fit a specific tissue defect [101]. In a medical application, a strength of bioprinting as a TE technology is that the 3D computer model of the final structure may be created from medical images. Medical imaging commonly uses a number of noninvasive modalities to obtain images that are used in clinical diagnoses and treatment. Computed tomography (CT), based on X-ray imaging, and magnetic resonance imaging (MRI), which uses nuclear magnetic resonance to form images, can both directly provide 3D data from which a computer model of a tissue defect specific to an individual injury might be generated and used in bioprinting a tissue construct [100, 102] (Fig. 1.4b). Other techniques which can provide personalized 3D representations of

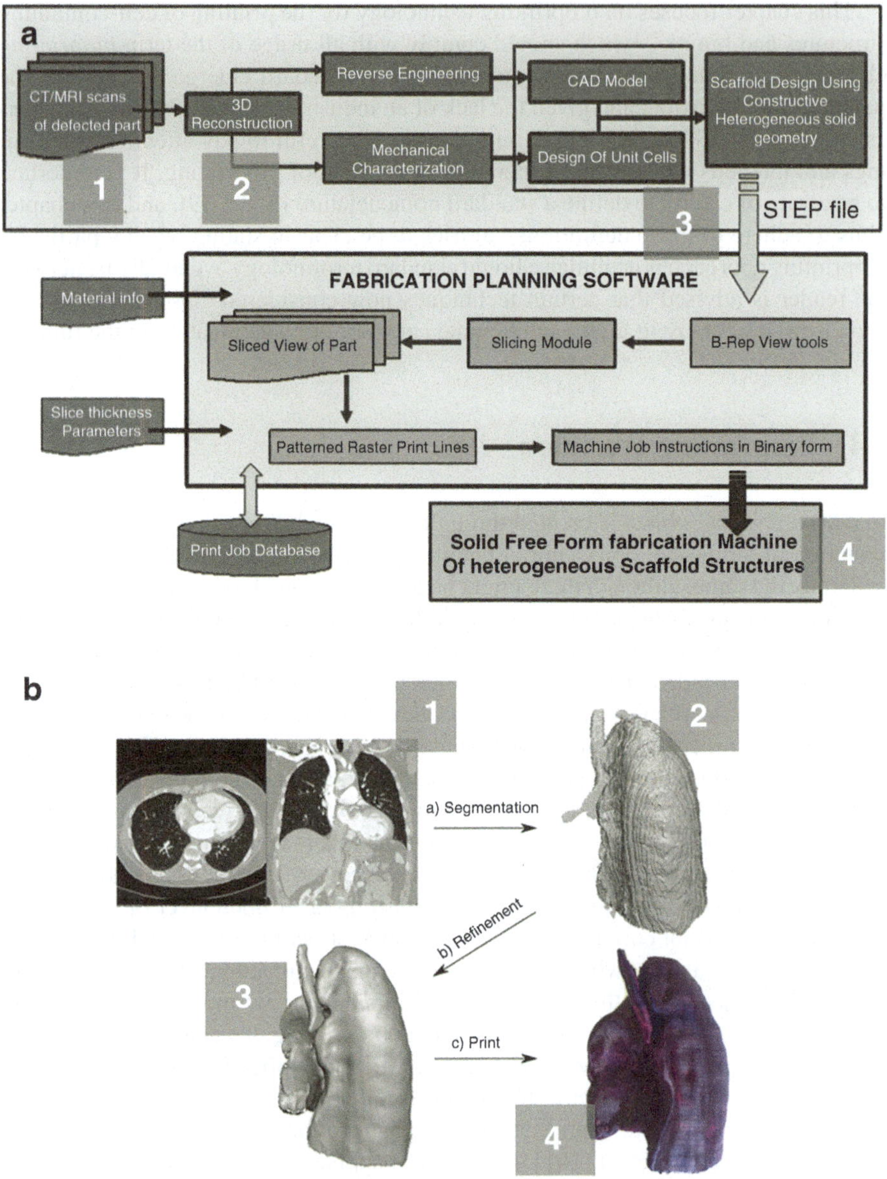

Fig. 1.4 (**a**) Flow diagram detailing the bioprinting process, from a medical image to printed construct—numbering corresponds to respective image in (**b**). (**b**) Images of the products (2D and 3D computer images and a printed construct) that might be seen at steps labeled by numbers in (**a**) ((**a**) adapted from Sun et al. Biotechnol Appl Biochem 2004 [55], (**b**) adapted from Bücking et al. PLoS One 2017 [100])

tissue structures include ultrasound and photoacoustic tomography (PAT) [103]. PAT uses a combination of optical excitation and ultrasonic detection to achieve extremely high resolution, multiscale imaging capabilities—it has been used to image blood flowing in individual capillaries [104]. These capabilities potentially complement multiscale capabilities available within bioprinting approaches.

Medical images might be used as the basis for the reproduction of a biological tissue, or they might be used to guide the design—as in injury or disease where original tissue is lost [65, 105]. The 3D computer model of the biological structure that derived from a patient or created de novo using CAD approaches can be saved in standard file formats that are easily shared between CAD software that include IGES and STEP. Often, computer models of objects to be (bio)printed using AM technologies are converted into a *standard tessellation language* (STL) format for printing. The STL file is then opened in software that computationally processes the object into layers (in a process referred to as slicing) and commands that control hardware movements. The STL file format was established with stereolithography printing in the 1980s, and it provides a digitized representation of surfaces of an object as being composed of discrete small triangles, based on the 3D coordinates of the triangles' vertices and vectors normal to their surfaces. An STL file specifies the shell of a volume of an object, but not the details of what is contained within the volume. An STL representation of an object can be shared easily, but the end user must know, or decide, what materials to use in printing that object.

With respect to processing of a digital 3D object into print commands, specific details may vary depending on the hardware or software used—particularly if proprietary or custom commercial systems are used. Specific print commands will vary with deposition methods as well, for example, continuous deposition of material in extrusion versus discrete deposition in droplet-on-demand, deposition methods versus light-based photocuring. Generally, the next step in the process would be importing the digital representation of the 3D object—the STL file, for example—into slicing software. According to software design and user-specified parameters, the object would next be computational "sliced" into thin layers. Layer thickness, and therefore z-resolution of the final object, would be controlled here. These layers would then be processed into a series of commands that will direct the printer's movements and material deposition. Again, depending on user input, many aspects of the hardware paths can be controlled, including important aspects of the process such as the rates of movement and extrusion and the spacing between printed paths—material spacing will correspond to print resolution and affect porosity. One can imagine bioprinting-specific slicing software that might automatically generate features such as voids that line up to form vessel-like channels.

The motion and extrusion commands that determine the print process are, at this point, a sequential series of movement commands, specific to that printer and that particular printed object. These commands are very often in G-code format—a format well established for computer-aided design and computer-aided manufacturing (CAD/CAM) processes (Fig. 1.5). G-code predates AM, having been used to direct numerical controlled milling machines in decades prior, for example. The sequence of print commands, whether in g-code or another format, can be saved as a text file,

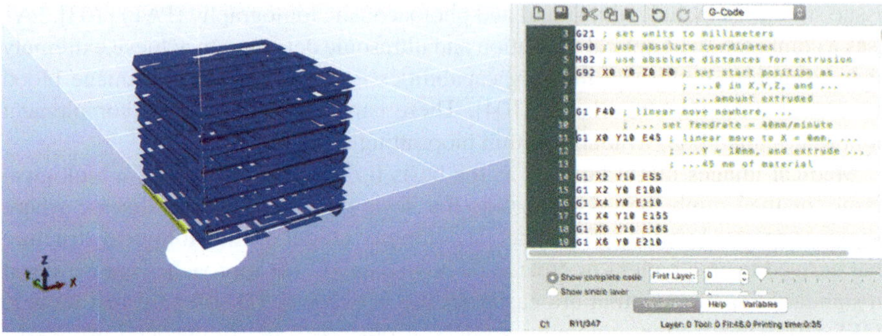

Fig. 1.5 Simple cube structure, sliced and translated to g-code to illustrate g-code commands. The yellow line in the bottom left of the image of the print paths on the left corresponds Line 11 in the g-code on the right (line is partially obscured by the white circle that represents the origin). This particular image is from Repetier-Host software. Complete lists of g-code commands are readily available online. Implementation of codes will depend on firmware

for example, for later use, reference, or even manual modification. However, it is the STL file that can most easily be shared, with each user slicing the object as desired and generating print commands specific to the printer they are using. Thus, one 3D model in STL format (e.g., a patient-specific tissue geometry) could be easily shared and fabricated on multiple printers—with perhaps more sophisticated systems offering increasing resolution or control over some other feature of the tissue construct.

In the final step of the (bio)printing process, the software that provides a computer interface with the printer sends the g-code (or similar) commands to the printer, where a microprocessor translates them into electronic signals that actuate the movements of motors and/or toggling of lights/switches/valves that comprise the physical printing process. This translation is mediated by firmware, a class of software that is stored on the hardware of a device—here the (bio)printer. In open source printers, firmware might be modified through a software environment on the computer, which can compile and upload modified firmware to the printer, enabling, for example, new g-code commands to be specified to yield new printing behaviors. In this way, g-code can be used to control printing functions in non-extrusion modalities such as laser-based sintering methods [106]. This general process will hold for many commercial printers as well, although commercial printers often use fully supported, proprietary software. There are always trade-offs to consider in choosing to use open source methods, particularly as commercially available printers become more affordable and increasingly sophisticated.

After a tissue construct is created, there might be post-processing to stabilize a printed structure [107–111] or remove support material [112–114]. Also, as in standard tissue engineering approaches, after fabrication of a structure, more steps may be necessary to create the final tissue construct. It may be necessary to introduce cells onto, or into, an acellular construct [115–118] for example. Time in culture may also be necessary for cells printed in a bioink to recover from the printing process, or to allow time for certain cellular processes to occur. For example, time may be required to allow cells to form endothelial layers within channels [119]. Some tissue constructs are designed to support the emergence of structure from the cells themselves, such as the formation of capillary networks [120, 121] or cell–cell interactions that result in tissue constructs whose function more closely recapitulates native physiological function [122].

In considering this process with respect to bioprinting specifically, as opposed to acellular printing, there are limitations in representing the full complexity of tissue using the standard approach of generating an STL, then slicing it, and finally generating print paths and commands. A macroscale structure represented by a single STL file might be designed to match a tissue defect, but without careful design of the process of slicing and generating print commands, the final bioprinted product might have the shape of a tissue or organ, but otherwise contain no more functionality than if the ink material were simply cast in a mold of that shape. Important physiological features, such as vessels and nerves, would be missing, as would microscale arrangements of cells. In particular bioprinting applications, this may not be important. However, when it is, slicing and path-generation software should be designed (or configured) to include porosity or to specify the printing of microscale structures such as vasculature, for example. If particular cellular or material arrangements are needed within a final structure specified using the standard STL approach, either multiple files describing the components of the structure would be needed, or the slicing and pathing software would need capabilities to allow the specification of such patterns within the part. The challenges surrounding achieving this are discussed at more length later in this chapter and in subsequent chapters in this book.

The process described and represents a basic workflow that would apply to standard AM as well as bioprinting methods. Bioprinting comes with additional considerations that range from how to prepare a biomaterial ink or bioink for printing and to load it into the printer. Maintaining a sterile environment is important if the printed construct will be incubated with mammalian cells or implanted in a patient. In the case of bioinks, sterility must be considered throughout their preparation and use. Furthermore, cells must be handled to avoid stimuli that might alter phenotypes, particularly in the case of stem cells, in undesirable ways or that might otherwise damage cells. Maintaining the viability of cells within bioinks throughout preparation, during, and after printing can be challenging. A further consideration that might result in deviation from the standard workflow is whether a printer must be modified in any way to facilitate the printing of a given material. Print processes may have to take into consideration the control of hardware that induces cross-linking in a bioink, for example, in order to maintain a stable 3D structure.

It should be noted that the general process of creating a CAD model, slicing it into layers, generating print commands, and sending these to a printer is not the only way to use a printer for bioprinting. Variations to the process might follow from unique hardware, software, or a given process itself. For example, printing need not always proceed in a layer-by-layer fashion, and the method for generating print commands would have to account for this. This might be the case in instances of specialized printer hardware, such as multi-axis robotic arms fitted with printheads and designed for printing on any surface from any direction [123]. Even using traditional printer technology, the printing of lattice structures—like those used to create channels within hydrogels from fugitive inks—might not be a layering process [124–127]. Instead, printer movements and material deposition might be continuous, moving as needed in any direction through three-dimensional space. If automated generation of print paths were desired, software would have to be designed to automatically determine these paths based on a given geometry—e.g., based on a desired 3D density of channels in the final volume. In the case of simple structures, print commands might simply be composed manually if a given software and hardware system allows for this. In the case of g-code, a simple series of commands can be easily written by hand and might be sufficient for initial testing of a new print process or the extrusion of a new bioink. At the end of the day, the (bio)printer is a tool that facilitates the controlled placement of (bio)inks in 3D space at high resolution. There is no single, correct way to use it. How the tool is applied to a given problem will depend on the researcher and her/his choices in adapting software and hardware to that application.

1.5 Bioprinting Technologies

Bioprinting technologies are diverse. Bioprinting has largely evolved from AM/SFF technologies, a number of which have been mentioned above. The field has also continued to evolve to include modifications to processes that are intended to facilitate achieving goals particular to bioprinting. As discussed previously, bioprinting aims to apply the strengths of SFF to addressing grand challenges in tissue engineering—in particular, fabricating tissue constructs with biomimetic complexity and function. Compared to other biofabrication technologies, the strengths that SFF uniquely brings together include high-resolution positioning capabilities (<100 μm) in 3D space; the ability to address these locations throughout a macroscale (<1 mm) volume in order to fabricate large objects; the repeatability and accuracy that comes from automated, computer-controlled fabrication hardware; the ease of redesigning and iterating on designs that come from CAD; the possibility to fabricate a patient-specific tissue construct; and the accessibility of the technology—at this point, these strengths are available in even the least expensive do-it-yourself printers.

Thus, through the application of SFF to TE in bioprinting processes, researchers can rapidly design and build scaffolds with multiscale architecture—including porosity at the mesoscale (100 μm to 1 mm) that is critical to the survival of cells within larger constructs. Through material choices, further features of a construct might be engineered—including biochemical and biomechanical cues that are important to directing cellular behavior or aiding in regeneration. Bioprinting offers capabilities to structure multiple materials with increasing complexity throughout the printed construct. Through the continued development of materials and methods for bioprinting, the use of bioinks also enables the controlled fabrication of scaffolds, or scaffold-free structures, with high cell densities and with control over the placement of heterogeneous populations of cells (i.e., multiple cell types) throughout their volumes. This capability is important in the fabrication of tissue- and organ-like structures, and it represents an important advance over traditional scaffold manufacture and seeding methods from tissue engineering and acellular bioprinting.

In reviews of the field, SFF and bioprinting technologies are typically described grouped into categories that organize the many varieties of the techniques into a few, common modalities. These categories typically include the three categories that will be used here: **extrusion-based methods**, **droplet-based methods**, and methods based on the application of light or heat, i.e., **energy-based methods**. Not all techniques can be easily categorized within a single category—for example, laser-induced forward transfer (LIFT) printing uses a laser (energy) to generate a droplet that is deposited on the print surface. A decision to categorize a technology, such as LIFT, in one category over another may be regarded as reflecting the interdisciplinary nature of the field, where authors' decisions are derived from experience in developing printer hardware and/or materials for use as inks. Here, LIFT is categorized as droplet-based printing, because the printing method is established on the hardware side, and ongoing considerations for the development of this method—like many methods—focus on the design and behaviors of the materials used as bioinks. In terms of materials, it is more closely related to other droplet-based methods than to energy-based methods.

1.5.1 Extrusion-Based Bioprinting

Extrusion-based methods represent the most common and accessible bioprinting modality [102, 128, 129]. Extrusion methods are characterized by the continuous extrusion of a bead of material—or filament—from a nozzle. In bioprinting applications, the (bio)ink to be printed is often contained in the barrel of a syringe to which a nozzle is attached. Flow might be driven by pneumatic pressure [130, 131] or through displacement of the material with a syringe plunger or a screw driver [67, 132, 133] (Fig. 1.6a), with mechanical extrusion offering potential advantages in flow control owing to lags in building up and relieving pressure in pneumatic approaches. The nozzle through which material is extruded is commonly a blunt syringe needle, a conical dispensing nozzle, or a glass capillary tube. Of course, more sophisticated hardware design is possible, and the engineering of printer components, such as the nozzles used in extrusion-based bioprinting, offers opportunities to expand capabilities in bioprinting.

Materials used in extrusion-based methods as (bio)inks must be capable of flow through the nozzle, where the cross-sectional geometry of the orifice has a large influence on filament geometry. Upon leaving the nozzle during extrusion, the material in the printed filament must rapidly transition to a stable state so that the three-dimensionality of the deposited filament is preserved, enabling a 3D structure to be built, layer upon layer. This transition is dependent on the properties of a given material, and so some extrusion-based methods are closely tied to a particular material or class of materials. A number of commonly defined modalities are listed

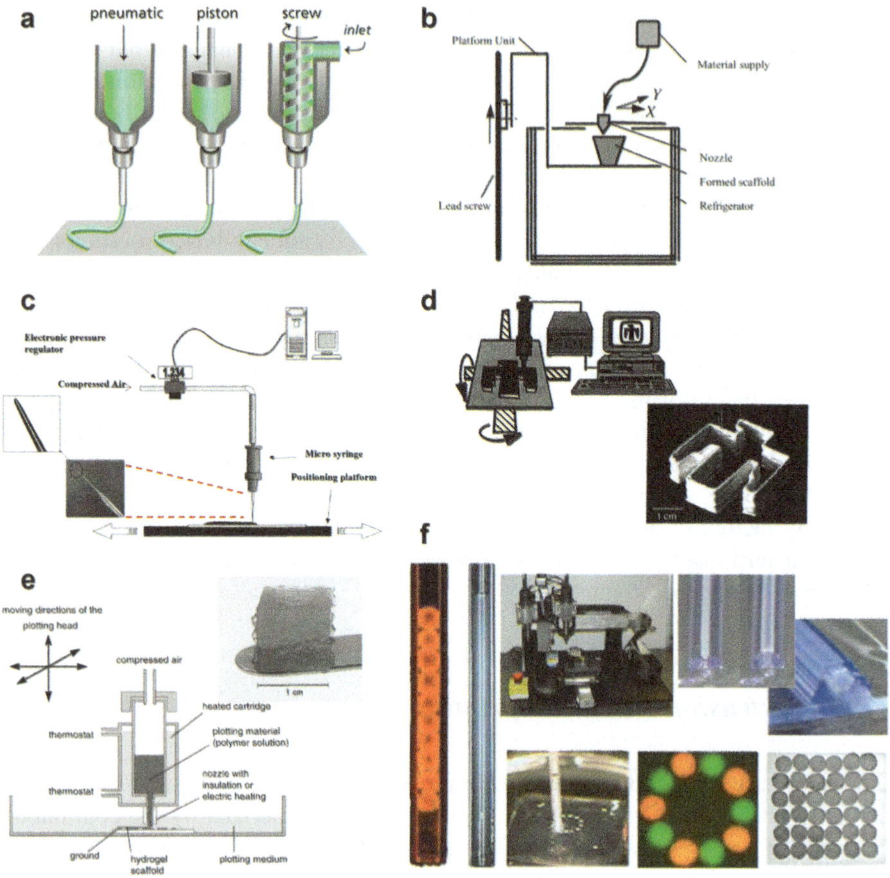

Fig. 1.6 Extrusion-based AM and bioprinting techniques. (**a**) Robotic dispensation. (**b**) Low-temperature deposition. (**c**) Pressure-assisted micro-syringe extrusion. (**d**) Robocasting. (**e**) Bioplotting. (**f**) Cell spheroid-based bioprinting, showing spheroids and continuous cell filaments in a syringe, printer hardware, and cell spheroids deposited in support gels and upon sacrificial support filaments ((**a**) adapted from Malda et al. Adv Mater 2013 [80], (**b**) adapted from Xiong et al. Scr Mater 2002 [134], (**c**) adapted from Vozzi et al. Tissue Eng 2002 [135], (**d**) adapted from Cesarano et al. 1997 [18], (**e**) adapted from Landers et al. Biomaterials 2002 [57], (**f**) adapted from Jakab et al. Biofabrication 2010 [136])

below, but most extrusion-based bioprinting using hydrogels and bioinks do not fit in one of these categories and would simply be grouped as "extrusion-based bioprinting" without further classification.

Thermal transitions are utilized in ***fused deposition modeling (FDM)*** (Fig. 1.1d). Given its long history and wide use in open source technologies, FDM is one of the most well-known well-established SFF methods. In FDM, thermoplastic material is driven through a nozzle that is heated to a high temperature at which the thermoplastic material behaves almost as a viscous liquid. These materials extrude as a fine filament (~200 μm diameter). These materials are entirely polymeric—in contrast to a hydrogel which is >90% water—they exhibit high viscosity at print temperatures, and rapid cool and solidify, and consequently are extremely well suited to SFF. However, the polymeric materials used in FDM and similar processes cannot easily be formulated as bioinks for bioprinting applications. Even in acellular bioprinting, the elevated temperatures required for extrusion often subject materials to temperatures approaching 100 °C and higher, which can be detrimental to the bioactivity of scaffold materials [137]. However, these have been used in composite inks that include ceramics and metals [92].

Cooling extruded materials can also induce thermal transitions that stabilize extruded biomaterial inks. In ***low temperature deposition modeling*** (LDM) [134] (Fig. 1.6b) and ***cryogenic 3D printing*** [138], for example, the printed acellular biomaterial inks were mostly solvent by weight fraction, and by controlling the temperature of extruded material to achieve freezing prior to flow of the solvent-rich inks, stable filaments could be printed in ways that preserved scaffold bioactivity. However, like FDM, the temperatures used were incompatible with bioinks, and thus printing was acellular. In order to harness thermal transitions in applications of bioprinting using bioinks, the temperature range over which the bioink must transition from flowing to stable must be cytocompatible, which means that it should be above 0 °C, as cells and bioinks are highly hydrated and ice formation must be avoided, and temperatures should not exceed 37 °C by much or for long periods of time, as elevated temperatures can be cytotoxic and cause proteins within cells to denature.

Other examples of acellular extrusion techniques used to fabrication tissue engineering scaffolds include ***robocasting*** and ***pressure-assisted micro-syringe (PAM) extrusion*** [135, 139]. PAM extrusion uses small, glass capillary-based nozzles to deposit materials dissolved in volatile organic inks [135] (Fig. 1.6c). The rapid evaporation of the solvent after printing leads to formation of a solid construct, with viscosity of the ink and solvent volatility being key features of the print process. In tissue engineering applications, scaffolds have been seeded with cells to provide structural, mechanical, and biochemical cues [139]. Robocasting specifically refers to the extrusion of ceramic slurries, although the printer hardware itself is similar, if not identical, to printer hardware that can be used in the extrusion of other materials that behave similarly to ceramic slurries in printing (Fig. 1.6d). Robocasting was developed as an alternative to traditional casting approaches for creating ceramic structures [18, 19] and instead used robotic dispensation (essentially, extrusion-based printing from a current view) to deposit ceramic inks. A key point of innovation in robocasting technology are the inks, whose rheological properties were remarkable in their ability to flow under shear stress and return to a solid-like state with a yield stress upon deposition. Printed structures might solidify further as water in the slurry evaporated.

In work following robocasting, colloidal gels exhibiting a similar viscoelastic response were designed to be printable materials in extrusion-based freeform fabrication [23] (Fig. 1.6a). Although they were not initially developed for bioprinting applications, materials like these—where, for example, particulate [23], polymeric [140], and/or ionic content [140] enable filament stabilization upon deposition—are noteworthy in discussing extrusion-based technologies. These materials were printable simply through a design that gave them inherent ability to shear thin and immediately set upon deposition. Filament sizes crossing many orders of magnitude have been printed, allowing fine and complex features [140], even dynamic [141], structures to be achieved according to nozzle and material design as well as print parameters. The extrusion-based process for printing inks formulated to establish stable filaments immediately upon leaving a nozzle has been described simply as a DIW process, but given no further classification (such as robocasting, FDM, bioplotting, etc.). Perhaps because of the early impact of these materials, DIW has become synonymous with the extrusion printing of colloidal gels, although DIW originally encompassed all the techniques discussed here as extrusion technologies, as well as droplet-based techniques discussed later [27].

Two extrusion-based technologies that were significant milestones in developing bioinks for bioprinting were ***bioplotting*** [57] and the ***printing of cell aggregates*** [58, 136, 142–145]. Bioplotting was the name given to the technique first developed for printing hydrogel structures. It was distinguished [146] from other extrusion approaches through a method of printing a hydrogel precursor solution into a liquid medium that was designed to support the formation of the hydrogel structure (Fig. 1.6e). The liquid medium's density and viscosity were key factors in successfully printing low viscosity aqueous solutions that could form hydrogels by a variety of mechanisms—thermal changes, a reaction with a coreactive component in the plotting medium, or multi-component dispensation. The flow could be driven pneumatically or mechanically, and material deposition could be in the form of continuous strands or microdots. While bioinks were not used in the initial work, cells could be seeded onto the scaffold after printing [57]. The printing of hydrogel materials containing cells is a dominant form of bioprinting today.

The printing of cell aggregates was essentially an extrusion-based technology that represented a prominent initial approach to using SFF-type technology for the purposes of fabricating tissues and, ambitiously, full organs. The distinguishing characteristic of this method was the use of cell aggregates as a biological ink—as mentioned before, the term bioink has roots in this technology. Though polymeric material can be included in the cell aggregates, its purpose is largely to prevent the merging of individual cell aggregates prior to printing. Cell aggregates are formed in advance, as can be achieved through hanging drop culture or non-adherent tissue culture dishes. These aggregates are then loaded into a microextrusion printer and deposited into positions on the print surface (Fig. 1.6f). Various hydrogel materials have been used to support the aggregates during fabrication—for example, printing into thin, intermediate layers of hydrogel [147] or onto cured hydrogel supports [143]. Microneedles have also been used as supports [148]. However, these materials are viewed as removable supports and not scaffolding in the traditional TE sense; hence, this is considered a scaffold-free biofabrication/TE technique. Minimal material exists between cells within these structures, as the intent is to facilitate self-assembly of dense cellular structures from the fusion of aggregates. The resulting, macroscopic construct would have cell densities approaching native tissue. The self-organization of the cells is intended to allow them to define cellular and extracellular structures, including the construct mechanics and functional properties.

Despite the many categories listed here and the expanding examples of extrusion-based bioprinting, most extrusion-based technologies used to print bioinks today might simply be best referred to as ***extrusion-based bioprinting*** methods without any further classification. That would also put these bioink extrusion technologies under broad definitions of DIW or robotic deposition. Because almost all methods for bioink extrusion process bioinks by moving them from a syringe to a build platform through a syringe needle or tapered conical nozzle, what distinguishes these techniques are the means for stabilizing the extruded materials. Stabilization methods are as diverse as the material systems used—materials used in bioprinting will be covered in more depth in the next chapter—attempting classification into categories specific to each system could yield a very large number of individual categories.

1.5.2 Droplet-Based Bioprinting

Droplet deposition techniques deposit discrete volumes of a biomaterial ink or bioink onto a surface by means of a variety of droplet-forming techniques. The movement of the material from the ink source (or printhead) to the print substrate is always in the form of droplets, but the means by which the droplets are formed varies depending on the technology used. Upon reaching the print surface, materials in the droplet must rapidly stabilize, prior to radially flowing and spreading on the surface, in order to establish a three-dimensional structure. The behavior of the droplet impacting the print surface is mainly influenced by velocity, (bio)ink viscosity, and the relative hydrophilicity of the droplet and surface materials [149] (Fig. 1.7a, b). Higher viscosities will lead to lower spreading but may be harder to print. Materials have been stabilized upon reaching a print surface in various ways [152]: by printing onto a surface that contains a cross-linking agent that can rapidly diffuse into the printed ink, by using an additional printhead to print a cross-linking agent directly over a bioink previously printed from another printhead, and by printing onto a surface which is being continuous exposed to an agent that induces cross-linking—for example, continuous exposure to UV light in the case of a photo-responsive ink or, as another example, to a mist of a fluid in which a chemical or ionic (in the case of alginate, especially) cross-linking agent is dissolved.

These behaviors must be balanced against material considerations in droplet formation (Fig. 1.7c). In particular for droplet-based methods, how materials affect a droplet forming at the ink source must be balanced against the requirement that they rapidly stabilize at a print surface. During droplet formation, surface tension and lower material viscosity are important in the process of a discrete volume of a (bio) ink breaking off from the source and forming a droplet. The forces imposed on the fluid during the process must overcome forces that keep the fluid together in order for droplet breakoff to occur. Many of the technologies are jetting technologies, which generate a jet of material which, through Rayleigh instability, breaks to generate a droplet. Droplet-based methods can be characterized as techniques by which

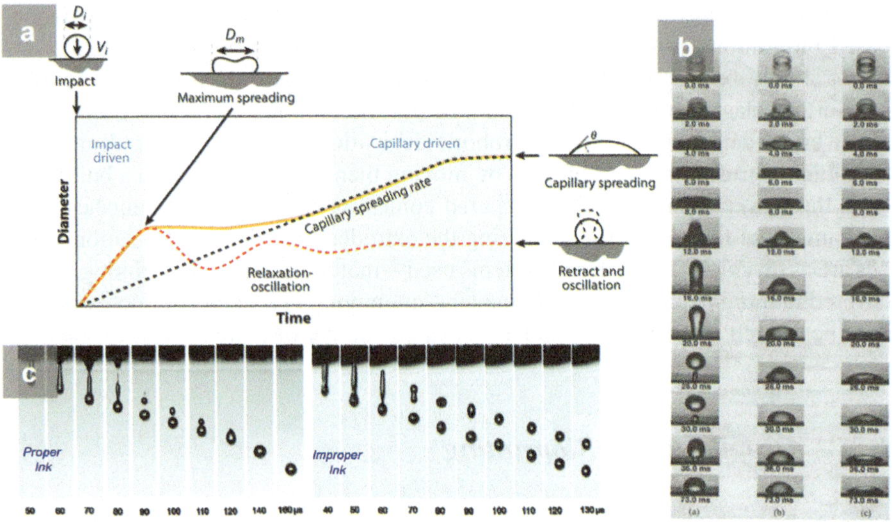

Fig. 1.7 (**a**) Droplet behavior upon impacting a surface—diameter varies with time as forces relating to impact and surface interactions act on the droplet. (**b**) High-speed images of a large (2.7 mm) aqueous droplet impacting surfaces of increasing hydrophilicity (left to right). (**c**) High-speed images of droplet formation, with two different ink formulations illustrating challenges in dynamic formation of a single, controlled droplet ((**a**) adapted from Derby Ann Rev Mater Res 2010 [150], (**b**) adapted from Mao et al. AICHE J 1997 [149], (**c**) adapted from Jang et al. Langmuir 2009 [151])

droplets are generated on demand or continuously [153]—control over droplet generation is hardware-dependent.

Resolution of a printed structure will depend on various factors that are dependent on both the materials used in the (bio)ink and the particular technology used for droplet generation [154, 155]. Droplet size affects maximum potential resolution, and it is related to both materials used and to droplet formation technology. How much a droplet spreads at the print surface in the x/y directions will affect the ability to achieve this resolution. Spreading is influenced by velocity of the droplet and material properties such as viscosity, interactions with the surface and other droplets at the surface, and rate of cross-linking (Fig. 1.7a, b). Resolution will also depend on control over the droplet generation process—whether droplet sizes are uniform and droplets can be precisely generated and placed according to a design.

Inkjet-based droplet formation technologies are well-known. Most people are familiar with the term from the technology's ubiquitous application in computer printers used to print documents. Among bioprinting technologies, as has been mentioned, inkjet printing has the longest history, having been used in the late 1980s to print bioinks [16]. It would be years before cell printing would be taken up again using any SFF modality. Inkjet printheads exist in different formats, for example, single- and multi-nozzle dispensers [128], and generate droplets with diameters on the order of 10s of microns, with corresponding volumes in the range of 1–100 pL

[150]. These droplets are generated by the ejection of a jet of the ink material from a small nozzle, the subsequent breaking off of a droplet. As with other methods, droplet breakoff is a function of physical features of the printhead as well as viscosity, surface tension, and density of the ink [151]. In inkjet printing, ejection of a droplet is triggered by a pressure pulse within a fluid-filled chamber in an inkjet printhead. This pressure pulse can be created using thermal [156–158] or piezoelectric (mechanical) [159–161] mechanisms (Fig. 1.2c). In the former, the pressure pulse is generated from the formation (and collapse) of a vapor bubble that results from heating a small pocket of fluid by a microheater. In the latter mechanism, a piezoelectric device creates a mechanical volume change resulting in the displacement of fluid [153]. Droplets are emitted through nozzles with very small orifice diameters, and in the absence of the actuating pressure pulse, surface tension at the orifice prevents (bio)ink materials from leaking out of the reservoir [150]. Pressure pulses must overcome this surface tension during droplet ejection. Since its initial application in bioprinting, inkjet technology has been used to deposit a variety of bioactive inks and cell-containing bioinks [95, 156, 157, 159, 162, 163].

Microvalve-based droplet formation technologies can also generate droplets on demand, but differ from inkjet methods in the hardware behind the droplet generation. In a microvalve approach, a (bio)ink is driven into the print nozzle by pneumatic pressure. The nozzle orifice can be opened or closed through the actuation of a small valve (the microvalve). This control is achieved via solenoids that open or close the valve in response to electric current [164, 165] that generates a magnetic field that acts on the plunger in the valve (Fig. 1.8a). Droplets are generated when the pneumatic pressure drives the (bio)ink out of an open valve, requiring that forces related to surface tension at the opening and viscosity are overcome. Microvalve actuation times are on the order of tenths of milliseconds at their shortest, corresponding to droplet generation rates of thousands of hertz [170]. A number of parameters thus impact droplet formation. As in other droplet methods [151, 171], chemistries within the bioinks (polymeric interactions or cross-linking) will affect the balance of fluid forces involved in droplet formation (pressure-driven flow vs. viscosity, surface tension, and capillary forces). In microvalve techniques, the time the valve is open can also be controlled as needed for droplet generation. As in other droplet-based methods, materials in the (bio)ink must rapidly stabilize to maintain a 3D structure. As with inkjet techniques, microvalve printing can be used with bioinks to print cell-containing structures [170, 172].

The use of ultrasound (>10 MHz) waves to generate droplets has demonstrated in work on ***acoustic bioprinting***. High frequency sound, or pressure, waves are generated using ultrasound transducers that focus these acoustic waves at a point near the surface of a fluid–air interface. The convergence of the acoustic waves at a focal point results in a localizer perturbation that can eject picoliter-sized droplets [167] that can include cells that are suspended in the fluid [168, 173] (Fig. 1.8c). Microfluidic devices with integrated ultrasound transducers [174] as well as ultrasound transducers facing upward toward the surface of fluid in a tank [168] have been demonstrated in the controlled generation of cell-containing droplets without pushing the bioink through a nozzle. Where the ultrasound transducer was oriented

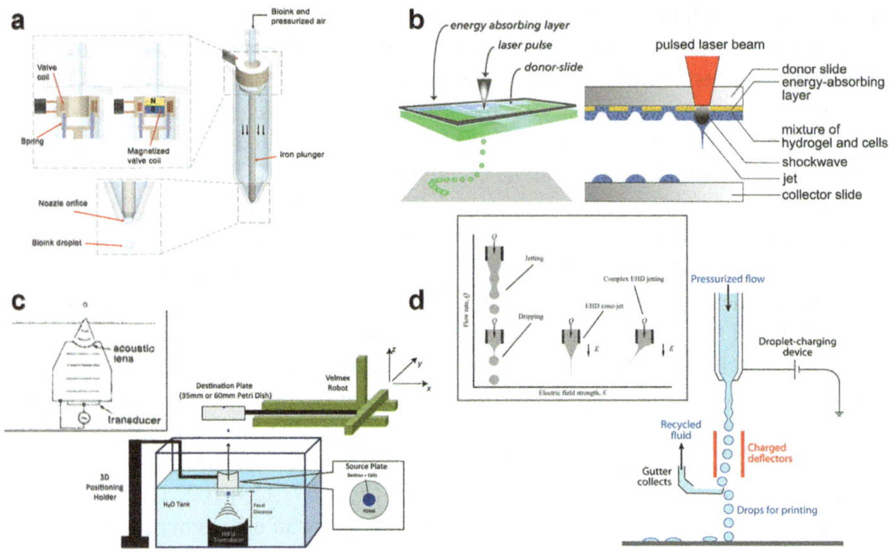

Fig. 1.8 Droplet-based AM and bioprinting techniques. (**a**) Microvalve droplet-on-demand hardware. (**b**) LIFT schematics. (**c**) Acoustic bioprinting schematics: inset, printhead only; in color, full "printer" system. (**d**) EHD (or continuous inkjetting) schematic (in color), with other potential jetting modes illustrated as a function of electric field strength and flow rate (in inset) ((**a**) adapted from Gudapati et al. Biomaterials 2016 [152], (**b**) adapted from Malda et al. Adv Mater 2013 [80] and Gruene et al. Biomed Eng Online 2011 [166], (**c**) adapted from Elrod et al. J Appl Phys 1989 [167] and Fang et al. Tissue Eng Part C Methods 2012 [168], (**d**) adapted from Derby Ann Rev Mater Res 2010 [150] and Onses et al. Small 2015 [169])

to focus beams upward, droplets could be ejected vertically from the tank. By modulating features of the acoustic waves—wavelength, amplitude, and duration, for example—droplets with diameters ranging from 3 to 200 μm could be created [174] at rates ranging from 1 to 10,000 Hz, with high cell viabilities.

In ***electrohydrodynamic*** (EHD) droplet generation, a high voltage is applied to a conductive nozzle, relative to a grounded print surface [175]. The potential creates an electric field between the nozzle and the surface and induces a charge on the fluid in the nozzle, similar to electrospinning [169, 176]. Droplet generation occurs continuously in this process. A (bio)ink solution is driven by low pressure into the charged nozzle, which induces charges within the solution. At the end of the nozzle, behavior of the material solution depends on the balance of flow rate and material composition—which dictates critical properties such as surface tension and viscosity, the voltage applied, the electric field (a function of factors such as voltage, nozzle distance to ground, and geometries of conductive surfaces), and electrostatic forces within the material [169]. Voltages applied in EHD droplet formation range from 0.5 to 20 kV [152, 177, 178], with voltages at the lower end of this range being demonstrated in bioink applications. Electrostatic forces within the material induced by the applied voltage result in repulsions that influence droplet breakoff. The balance of these factors determines whether a given solution leaves the nozzle as a stream of discrete droplets, which is the desired mode in EHD-based printing—as opposed to, for example, larger drips, jets, or fibers [153, 169, 175] (Fig. 1.8d). To control the droplets reaching a surface, the electric field they pass through during flight can be altered—because the droplets are electrically charged, they can be diverted into a collector gutter for recirculation if they are not desired at the print surface [153]. In a bioink, the material will contain cells, and high cell viabilities have been reported after EHD processing [175, 177, 179–182].

A drop-on-demand technology that uses laser light to control droplet generation falls into a gray area in terms of categorization as either a droplet-based or energy-based method. ***Laser-induced forward transfer*** (LIFT) bioprinting involves the transfer of a (bio)ink as a droplet from a LIFT-specific substrate to a print surface upon application of laser light. In terms of material considerations, the method is droplet-based. In terms of the printer hardware, it is an energy-based method. Key components of the printer include the laser, an ink-coated substrate (the ribbon or donor slide), and the print surface (the print substrate or collector slide) [183–185]. The donor slide is coated with a layer (10s of nanometers) of a metal or metal oxide that absorbs the laser energy (Fig. 1.8b). A layer of material to be printed—for example, a cell-containing hydrogel—is homogenously coated on top of the laser-energy-absorbing layer. The hydrogel layer is typically 10s of microns thick, and multiple regions of hydrogel may be coated onto the slide if multiple formulations of (bio)inks are desired (e.g., multiple cell types). In the printing process, this donor slide is positioned so that the coated side faces the collector slide, and laser pulses are directed through the back of the glass donor slide to the energy-absorbing layer. There, at a spot size of 10s of microns in diameter [183, 184, 186], the energy absorbing layer melts within picoseconds and then vaporizes [166], creating a pressure pulse in all directions—but with the least resistance away from the slide, as the thickness of the coated bioink is thin there, compared to that in the lateral directions [166]. The resulting jet of material [187, 188] will result in a droplet that travels to the collector slide which is located hundreds of microns below the donor slide [184]. This technique requires no nozzle and so can be used with high cell densities without concern for clogging a nozzle. The collecting substrate typically has a thin hydrogel or culture medium layer to cushion the impact and help hold cells in place [184]. The resolution achieved depends on the material properties, particularly viscosity and relative hydrophobicity on the print surface—as in other droplet-based bioprinting methods. Particular to LIFT, the laser energy, pulse time, and the thickness of the (bio)ink layer on the donor slide will also affect droplet sizes and resolutions achieved. Bioprinting has been demonstrated with multiple cell types and at high cell densities [185].

1.5.3 Energy-Based Bioprinting

Methods included in this category are characterized by the focusing of an energy source, often a laser or high energy light source, at the print surface where the material to be added to the printed construct is already present. Exposure to that energy—heat or light in the methods discussed here—results in the solidification or stabilization of the material. In most of these methods, the material must be constantly replenished as one layer is stabilized and the next is built upon it, so there are considerations of flow and transport in material design. However, the nature of the (bio)ink transport differs from extrusion- and droplet-based methods in that materials used in energy-based methods are cured/set/cross-linked in place, rather than moved as a discrete filament or droplets from a printhead to the build surface. As in the methods discussed in the previous sections, fabrication using energy-based approaches usually occurs in a layer-by-layer manner. In contrast to those methods, the first step requires introducing material to the entire build surface. Next, an energy source is directed at the build surface, focused at a single location, or pixel, and then scanned across the material, point-by-point (Fig. 1.2b). Alternatively, in some cases, the whole layer can be exposed at a single time to a projection of the curing energy—typically a light source that is inducing photo-chemical curing/cross-linking in cases of projection. Whether the energy is scanned (or rastered) across the surface point-by-point, or projected across the entire layer, upon completing the layer, the build surface typically moves away from the energy source by a single step in the vertical direction, with the step size determining z-resolution. New material either fills in the space left by the previous layer—as in the case of techniques where curing occurs in a vat of a liquid material that can flow into place—or is applied by the printer—as in the case of powder materials that are pushed into the place in certain processes or in the case of liquid materials that are introduced under pressure-driven flow. This repeats until the object being fabricated is complete.

Selective laser sintering and closely related selective laser melting [189] fuse or melt together powder, or particulate, materials using thermal energy from a focused laser beam. The laser beam is scanned as described above, with successive layers of power typically from 20 to 150 μm thick, depending on the printer [190]. In terms of forming structure by selectively binding together powder precursor, this method has some similarities to the printing of a binder material onto a print bed containing a powder that is chemically fused, as in the case of the method to which the term *3D printing* [31] previously referred to exclusively (Fig. 1.9a). These powder-based methods are inherently acellular—the processes use energy or chemistry that would kill cells, and the environment is not aqueous, but dry, which would desiccate cells—and thus they can be used in bioprinting primarily in fabricating scaffolds for tissue engineering or regenerative medicine. Materials used must be formulated as powders with particle sizes on the order of 10s of microns. In SLS printing, the powder materials must melt, sinter, or be coated such that they will bind together during heating, and they must flow [92]. Open source hardware has recently been developed [106] and—in a unique application of SLS technology—used to print removable poly(caprolactone) structures that could be embedded in poly(dimethyl siloxane) and then dissolved away with a solvent to leave behind channels through the elastomeric structure.

Compared to the many methods—in extrusion- and droplet-based bioprinting especially—in which bioinks can be printed to fabricate cell-containing structures, ***stereolithography (SLA)***, the technique that was granted the first AM patent, has taken longer to develop [87]. Challenges in developing this method of AM for bioprinting using bioinks will be discussed in the next section. The process itself is similar to SLS, in that a focused laser light (often ultraviolet light) is scanned across a print surface in the *xy*-plane, with the ink material that is exposed to the light being solidified according to the design for that layer. The build platform then moves in the *z*-direction, and new ink material flows into place at the previous *z*-location, and the process repeats [191] (Fig. 1.9b). The *z*-height of each layer is typically 25–100 µm. Often the distance of the *z*-movement of the build platform is less than the depth at which the laser light initiates curing/cross-linking, which allows for binding of material from one layer to the next.

Depending on the setup of the SLA hardware, the build platform can move either up or down—it always will move away from the laser source [191]. When the platform moves down, away from light coming from above, then new material must deposit at the top of the layer that was just cured. The platform may be immersing into a larger, surrounding vat of material or, if the build platform itself has side walls, the material might be added on top of the previous material (Fig. 1.9b). The former setup may require a large volume of uncross-linked material and would limit heterogeneity of material within the structure, whereas the latter setup might require extra steps or controls to implement. The top layer of the material which is being cured is often exposed to air in this setup. This approach has been referred to as a "bottom-up" setup, because the bottom layer is built first, and as the platform moves away from the laser, new, higher layers are built [24, 191].

In an alternative setup, the light initiating the curing or cross-linking can be incident on the material to be cured through a photo-permeable window, below the build platform. There would be a small space between the build platform and window on the first layer that would be filled with material to be cross-linked. Photo cross-linking must be carefully controlled to ensure that the layer that is cross-linked binds to the downward-oriented platform—and not the window. After each layer is complete, the platform moves upward, and new material flows into the small space. This process minimizes the amount of material needed per layer and can eliminate exposure to air (which can interfere with curing), but materials used must have rheological properties that match the requirements of flowing into the small space between window and previously cured layer [24, 191]. This method is referred to as "top-down" because the top layer of the structure is attached to the build platform and layers below it are cured as the platform moves upward, away from the light source. Of course, this nomenclature does not take into account the orientation of the product being fabricated, which might be inverted so that the bottom of the structure is attached to the platform and cured first—in which case, the process still may be starting at the bottom, from the printed construct's inverted frame of reference. ***Micro-SLA (µ-SLA)*** technology is functionally the same as SLA, but at a higher resolution and with a correspondingly smaller final construct size [24, 193].

Digital light processing (DLP) or digital light projection methods are very similar to SLA approaches in terms of general processing, with the key difference in how light is applied to the layer being cross-linked. In SLA methods, a laser is scanned, point-by-point, across an entire print surface to cure material where

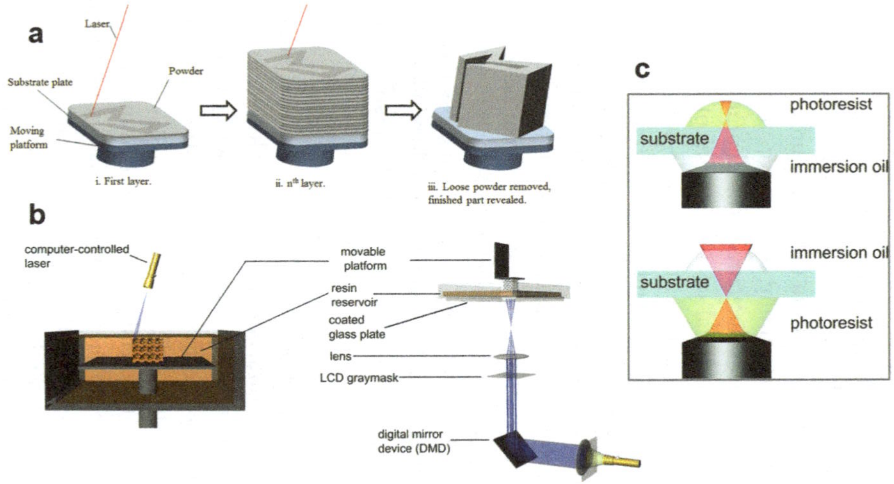

Fig. 1.9 Energy-based AM and bioprinting techniques. (**a**) SLS fabrication. (**b**) SLA (left) and DLP (right) with a stages illustrated that move either down into a resin reservoir (right) or up, pulling the structure out of the reservoir (right). (**c**) 2PP, either from beneath a slide, with light focused through the glass, into the resin to build a structure up from the glass, or with light coming from the same side as the resin—also building structure away from the glass, but avoiding focal length limitations on *z*-dimensions ((**a**) adapted from Yap et al. Appl Phys Rev 2015 [189], (**b**) adapted from Melchels et al. Biomaterials 2010 [191], (**c**) adapted from Bückmann Adv Mater 2012 [192])

desired. In contrast, in DLP, light is projected onto the surface of the entire layer at once, only illuminating regions that must be cross-linked [194] (Fig. 1.9b). Despite being called DLP and being differentiated from stereolithographic printing—the DLP approach is similar to photomask-based lithographic techniques such as those in semiconductor fabrication. With respect to mask-based lithography, the most outstanding difference in DLP is the digitally controlled projection of light onto the surface of the material to be cured. Compared to SLA approaches, the photocuring of an entire layer at once offers significant advantages in terms of processing time. Digital micromirror devices (DMD) and LCD projectors have both been used to project light onto a surface [195], with DMDs offering better performance in terms of transmitting short wavelength lights and avoiding challenges inherent in pixel-based LCDs, such as optical fill factor and switching speed [196, 197]. In DLP, as well as in SLA, in addition to material formulation, exposure times, layer thicknesses, and light intensity will all affect cross-linking of the material used (often called a resin, or bioresin [87, 198], in the case of SLA and DLP techniques). Recently, the speed of DLP-based cross-linking of an entire layer of material was combined with careful printer design to control where photo cross-linking of material occurred, to enable the layer-by-layer process to occur during continuous movement in the z-direction, speeding up print times immensely in a process dubbed "continuous liquid interface production" or CLIP [199]. While the CLIP process has not been implemented with bioinks, recent advances in materials have allowed for DLP-based bioprinting of bioinks [198, 200].

At the smallest length scales, ***two-photon polymerization (2PP)*** is a technology that can be used in biofabrication similar, in many ways, to SLA-based bioprinting [192, 201, 202] and therefore is often considered among bioprinting technologies—it is sometimes also referred to as *two-photon direct laser writing* or *two-photon lithography* or similar. It uses light to photo cross-link materials and, in terms of building structures, is often used with acellular biomaterial inks. However, in biofabrication approaches that are not additive—and therefore would not be considered bioprinting—it has been demonstrated in applications where it can modify features within light-reactive, cell-containing hydrogels to establish heterogeneity at the microscale [203–206]. However, it requires unique equipment and therefore is not broadly accessible. 2PP is based on two-photon absorption, a phenomenon that occurs when a molecule that can be excited by a photon at a given wavelength simultaneously absorbs two photons at a much longer wavelength (so with lower energy) to achieve the light-activated energetic state that the molecule would have otherwise only reached with the higher energy (shorter wavelength) photon. The probability of a molecule absorbing two photons on the incredible short—practically instantaneous—time scale required to achieve two-photon excitation is very small, and therefore, a specialized, femtosecond laser is usually used to deliver light in photon-rich pulses to focal point in 3D space [207] (Fig. 1.9c). In the case of 2PP, the molecules that are excited are typically photoinitiators that trigger the polymerization of the hydrogel precursor solution in which they are dispersed. At the point of focus, the photoinitiators will typically generate free radicals that locally react with the polymeric material in solution to initiate cross-linking at only that point

[207]. It is possible to achieve light-based patterning at a sub-diffraction limit resolution using this technology [208, 209].

Because the photons emitted from the laser are at much longer wavelengths (lower energy) than the photons required to initiate the reaction, the highly improbable two-photon absorption can only happen at the focal point, and any material that long wavelength photons travel through to get to that point remains unaffected. Thus, the focal point can be located anywhere within a volume with no worry about inducing photo reactions in the surrounding material, which is central to its usage in modifying cell-containing materials [203–206]. This is in contrast to standard light-based processes like SLA or DLP, where the light incident upon the (bio)ink is at the wavelength necessary to initiate a reaction, and therefore can cause cross-linking out of the plane of interest. As mentioned, in terms of printing, 2PP produces structures that might look similar to those produced using SLA techniques, but orders of magnitude smaller [192, 210]. Printing generally uses acellular biomaterial inks, with printed structures often designed to probe single-cell behaviors [202] or for in vitro studies of how cells interact with microscale structures [201, 211–213]. Because of limitations on how deep light can travel into a sample, the scalability of this approach is limited. While it has tremendous resolution at the microscale, macroscale structures are limited to millimeters in size.

1.6 Printing with Bioinks: Strengths and Challenges Within Bioprinting Technologies

Among bioprinting technologies are some shared strengths but also limitations related to common features of bioprinting processes. These shared strengths have been discussed and include hardware capable of high-resolution positioning in 3D space and the ability to fabricate—according to a computer design—a patient-specific or custom structure. In order to realize the full potential of bioprinting technology, challenges must be overcome. Within these technologies, there is a shared requirement of controlling transport and deposition (or solidification) of material, including cells in the case of bioprinting, into an organized structure—this lies at the heart of any bioprinting, SFF, or biofabrication process. Challenges surrounding the flow and stabilization of printed bioinks are a common theme and are addressed below with respect to specific printing technologies (Fig. 1.10a). Material choice or design, which is inseparably tied to this challenge, is addressed in a subsequent chapter.

Printing also, with a few exceptions, proceeds in a layer-by-layer manner, which means that fabricating a structure that extends continuously across z-space must occur through the discrete deposition of material defining that structure on each layer as it is built. Achieving complex, heterogeneous structures through 3D space, or even at high resolution in a single layer is also not a trivial consideration. The resolution of deposited structures is typically not as fine as the positioning capabilities of bioprinter. Achieving higher spatial resolution comes with the trade-off of increasing print times [99] and challenges associated with printing cell-containing bioinks. The

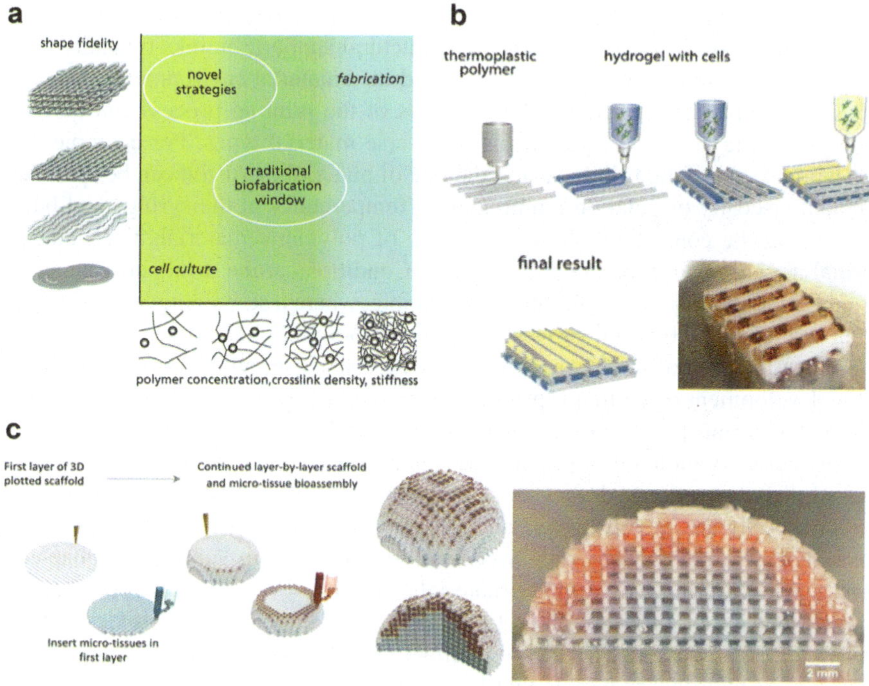

Fig. 1.10 (**a**) Schematic illustration representing a central challenge in bioprinting—materials used in fabrication have high polymer content and high shape fidelity; materials used in cell culture have low polymer content and cannot maintain shape; in bioprinting, we want to maintain shape with low-polymer content hydrogel-type materials. (**b**) Supported printing of soft hydrogel materials. (**c**) Supported printing of cell spheroids. In both cases with spatial definition of heterogeneous structure ((**a**) adapted from Malda et al. Adv Mater 2013 [80], (**b**) adapted from Schuurman et al. Biofabrication 2011 [214], (**c**) adapted from Mekhileri et al. Biofabrication 2018 [215])

layer-by-layer fabrication processes preclude the printing in arbitrary space—once material is deposited, a position below it cannot be altered, nor can material be printed in a location below which no support exists. Toward the goal of fabricating functional tissue structures using bioprinting, a given method must address challenges such as these and potentially others. Below, we consider strengths and challenges within bioprinting modalities, with a focus on bioink deposition. In the next section of the chapter, we will focus on innovation to bioprinting processes that are advancing capabilities but that cannot simply be associated with a single printing modality.

1.6.1 Considerations in Extrusion-Based Bioink Deposition

Extrusion-based bioprinting is widely employed in the deposition of bioinks—either as cell-containing hydrogels or completely cell-based (scaffold-free) type approaches. Extrusion-based bioprinting methods are compatible with many biomaterials that have viscosities ranging from 30 mPa·s to over 6×10^7 mPa·s [83]. While high viscosity materials offer good structural support, bioinks typically have lower viscosities that correspond to highly hydrated material environments that support cell viability and function. Many aspects of the printing process can be controlled to fit the printing requirements of diverse material types. For example, the translation speed of the printhead and the rate of material extrusion can be specified to match viscosity or gelation requirements. Temperatures at the syringe and build platform can be controlled, allowing the use of polymeric materials that undergo thermal transitions in biocompatible ranges. Conditions at the print surface can be varied—as in the case of bioplotting, for example, where a stabilizing liquid support might be present at the print surface into which bioinks could be directly deposited.

As mentioned, challenges surrounding the design of printable bioinks are focuses of the development of all major printing methods [80]. Bioinks are most often soft hydrogel materials [216], where generally less than 10% of the mass of the ink is in the polymeric component(s). When cross-linked, the polymeric components form a network that gives the hydrogel its stability and ability to resist deformation, but these materials are mostly water, and prior to cross-linking their viscosities are generally low. Specific materials used will be discussed in subsequent chapters. Upon leaving a print nozzle, these materials can flow away from not just their *xy* location, but importantly flatten and lose their three-dimensionality, unless cross-linking (or gelation) can be triggered almost instantaneously. Outside of a few of "workhorse" hydrogel materials that have fast cross-linking kinetics, like alginate and gelatin, modifications to materials or processing offer opportunities for expanding printable bioinks. Stabilization of polymeric networks within hydrogel materials prior to printing, for example by limited cross-linking [109] or physical bonds within the bioink [217], can yield materials which can flow during printing but achieve viscosities that resist flow immediately upon deposition. Using a photopermeable nozzle to induce gelation of materials that undergo photo cross-linking during extrusion is an example of how the process might be modified [218].

Simply increasing the viscosity of a bioink to help slow flow upon deposition must be considered against another major challenge in extrusion-based bioprinting: shear within the nozzle. The shear forces experienced within a material, and that are detrimental to the viability of cells within a bioink [219, 220], increase with dispensing pressure [221] and nozzle length and with decreasing nozzle diameter [219]. But low pressure and larger nozzle diameters come at the cost of print times and resolution. Materials which have, or are engineered to have, shear thinning (reduced viscosity during exposure to forces such as those experienced during extrusion) or thixotropic properties (reduced viscosity over time during exposure to stress and, conversely, increasing viscosity and stability—over time—after exposure) are important [222]. In terms of thixotropic materials, the timescales over which they physically "set" are an important consideration for maintaining structure of a printed filament.

Compared to other approaches, particularly droplet-based bioprinting, extrusion methods where filament is continuously extruded from a nozzle face challenges in printing heterogeneity within a structure with subtle changes from one location in

3D space to the next [164]. Because these methods are not designed for rapidly starting and stopping flow to deliver small volumes of material, subtle changes throughout a structure might be introduced through nozzles that can combine multiple inks [223], but printing heterogeneity on a point-by-point basis remains technically challenging and remains a hurdle in achieving cellular and material complexity at biomimetic resolutions.

Currently, most bioprinting methods—extrusion-, droplet-, and energy-based—share the challenge of achieving physiological cell densities. In the development of extrusion-based bioinks, cells are often included at densities on the order of magnitude of 10^6–10^7 cells/mL—generally a couple of orders of magnitude below native cell densities in tissues and organs [224]. Toward this goal, ***bioprinting using cell aggregates*** has the unique strength of using cell-based inks that contain minimal amounts of non-cellular material, if any, within bioink formulations. Dissolved polymeric components can be included, but generally to prevent aggregation of clusters. The structures printed from cell aggregate bioinks are scaffold-free and naturally have high cell densities, given that the bioink is composed of clusters of cells that serve as units of printable material. Though they are placed using an extrusion-based process, cell clusters are typically deposited discretely, like droplet-based printing. This allows potential for point-by-point control of heterogeneity in the structure, at the resolution of cell-cluster sizes. The extrusion of cell cluster-based bioinks may avoid some of the risks of damage from shear seen in more viscous, hydrogel-based bioink formulations for a number of reasons. The material the clusters are suspended in is typically not needed for structure after printing and therefore has low polymer content and consequently low viscosity. Flow within the nozzle that is likely plug-like with therefore low or no shear anywhere but at the wall of the nozzle, and cell–cell interactions may provide support that stabilizes individual cells against shear-induced elongation.

One challenge that all bioprinting technologies face, but that high-density cell aggregate-based printing highlights, is cell sourcing. The expansion and culture of clinically relevant numbers of cells to be used in bioinks, and the control of cellular phenotypes, is a significant challenge and a major area of research that is necessary to translating any cell-based technology to the clinic [225–227]. This is also a concern where progenitor or stem cells are used and predictable, standardized behaviors are required. Forming cells into uniform spheroids or aggregates is also a nontrivial processing step that is necessary in aggregate-based printing. Control over the ultimate composition of an aggregate-based construct is, by definition of printing with aggregates, not controlled at a single-cell level—the resolution of these structures depends on the size of the cell aggregates and usually is on the order of 100s of microns. Toward organizing into functional structures, these approaches rely on the cells themselves organizing and forming biologically functional structures [136]. Allowing for, and even depending on, cells autonomously self-organizing to give rise to function is not unique to spheroid-based approaches and in fact is central to TE approaches, going back to the earliest studies [228]. Cell-aggregate approaches simply do not have material scaffolding that might potentially aid in this process. The lack of material scaffolding also means that these approaches require some sort of support during and after printing, until aggregates fuse into tissue-like structures [142, 143] (Fig. 1.6f). Preventing this fusion during the printing process and achieving adequate transport of nutrients and wastes exchange within cell-dense structures, where nutrient consumption will increase with cell numbers, must also be considered.

1.6.2 Considerations in Droplet-Based Bioink Deposition

Droplet-based methods offer the potential to print a variety of material types: polymeric solutions, colloidal suspensions, and cell-containing bioinks. Droplets with diameters on the order of 20–60 μm and very small volumes (1–100 pL) [39, 96, 152, 229, 230] can be delivered with high positioning accuracy (~10 μm) [150, 153]. These capabilities offer the potential for precise control over spatial heterogeneity within a construct through the controlled deposition of these small droplets according a pre-defined pattern, as has been nicely demonstrated with acellular biomaterial inks [115]. Compared to non-droplet methods, this represents a significant strength inherent to droplet-based technology [164].

These methods employ materials with relatively low viscosities—compared to the range possible in extrusion-based printing—that are typically less than 10 mPa·s at high shear rates (10^5–10^6 s^{-1}) [92]. Formulating materials, especially bioinks, can therefore be challenging in droplet-based printing, as in the other bioprinting methods discussed, because of the competing requirements of flow and stability. In order to bioprint a 3D cell-containing structure using droplet-based methods, bioinks must first allow droplet formation, then maintain some three-dimensionality upon reaching the print surface where droplet motion toward the surface is stopped and the material begins to interact with the substrate, where it must then—almost simultaneously—rapidly stabilize/cross-link [150, 229]. Constraints on the materials that can be used in bioinks for droplet-based printing are therefore significant, even compared to extrusion-based methods, which is why, for example, viscosity ranges discussed below are more constrained in droplet-based approaches.

Inkjet-based bioprinting offers an accessible option to droplet on demand bioprinting through the possibility of modifying 2D inkjet printers [102] as well as purchasing printheads to build custom equipment [128]. Inkjet printheads offer high rates for droplet generation, up to 30 kHz [229]. Combinatorial delivery of materials, potentially containing cells and/or biochemical components, can be achieved from multinozzle technology that was originally developed for printing graphics using three primary colored inks plus black [153]. Despite some concern for thermal stress [128], the localized heating associated with thermal inkjetting leads to only small increases in the overall temperature of ink material within a printhead [158], and it has not had a significant impact on the printing of biochemical molecules [231], proteins [115], or cells in bioink formulations [156, 157]. In the case of piezoelectric printheads, concern has been raised about the frequencies used and their ability to disrupt cell membranes [232]; however, there is an advantage of avoiding temperature increases by using this printhead, and bioinks have been successfully printed using piezoelectric inkjetting [159, 171], including in complex 3D structures [160] and with attention paid to cell viability under a range of operating conditions [159]. Sedimentation of cells in the printhead while sitting within a nonviscous bioink has also been addressed with additives designed to create neutrally buoyant systems [171].

Inkjetting technology does use nozzles with small diameter outlets, and these can clog when printing viscous materials or if cells settle during the printing of a bioink [164]. Printable inks have viscosities in the range of 3–30 mPa·s [229], and at 10 mPa·s and above, significant forces are required to eject droplets, which may be detrimental to cell viability, putting the highest viscosities at the lower end of what can be printed using extrusion-based methods, for comparison. This puts limits on the types of materials that can be used and the cell densities that can be achieved in a bioink. The deposition of cells in bioinks by jet-based ejection of a bioink through a nozzle inherently will involve shear forces; however, careful consideration of the system can limit the forces to which cells are exposed [233]. As a final consideration, despite high frequency capabilities, droplet formation can be destabilized at elevated droplet generation frequencies, resulting in small secondary droplets that lead to variance in droplet size [234, 235], resulting in recommended droplet generation rates that are significantly lower than 30 kHz.

The use of ***microvalve-based bioprinting*** to generate droplets on demand shares key droplet-derived strengths with inkjet-based technologies. In particular, this technology has been used to position (or deposit) cells to develop precisely controlled multicellular structures with high viability [236]. Compared to inkjet printing, the nozzle size is a little larger, and materials with a wider range of viscosities can be printed, ranging from 1 to 100 mPa·s. Higher viscosities may be advantageous fabricating 3D structures, as they will absorb more energy and spread less when bioink droplets impact a print surface [164]. The maximum rate at which droplets can be generated is ~1 kHz, which is therefore significantly lower than the maximum rate that can be achieved by inkjet printing. However, droplets are stable at this rate with no secondary droplets breaking off as the material is jetted out of the nozzle—consequently, in bioprinting, microvalve technologies compare favorably to inkjet approaches in terms of droplet generation rates [170, 235]. As a tradeoff for these strengths, microvalve-based droplet generation does result in larger volume droplets [237], and therefore decreased resolution, compared to other droplet-based methods. Despite having a larger nozzle orifice, cell sedimentation can still result in clogging, and therefore cell concentrations in bioinks are typically lower ($<10^6$ cells/mL) [170, 238]. While more viscous materials can be used, relative to inkjet bioprinting, compared to most other bioprinting technologies, the viscosity range is still restricted. Like other techniques (inkjet-based bioprinting and extrusion-based bioprinting), shear experienced by cells traveling through a narrow nozzle is a concern that must be addressed through print parameters, hardware design, and materials engineering [233].

Acoustic bioprinting offers notable strengths in terms of eliminating a number of potential cell-damaging features of other droplet-on-demand processes. By using focused sound waves to generate droplets as opposed to piezoelectric or thermal actuators, concerns relating to heat, pressure, and frequency that are present in inkjet methods are avoided. Droplet formation occurs where ultrasound is focused, without the need for a nozzle, and thus nozzle clogging and shear stresses that might be detrimental to cell viability in bioinks are not present when working with this technology [168, 173]. Acoustic bioprinting of bioinks has demonstrated its potential to create complex cellular structures, but the technology is not widespread, which means equipment is not easily available and must be custom made, requiring some experience with the technology. It is likely that, as in the other jetting techniques discussed, upper limits on viscosities will be low relative to extrusion-based technologies to enable droplet formation.

Electrohydrodynamic droplet generation offers advantages in being able to work with solutions that, relative to other droplet-based methods, contain high concentrations of dissolved material. Because of the way droplets are formed, this process can achieve very small droplet sizes, with diameters smaller than that of the nozzle from which they are being emitted [175, 182]. Based on process parameters, particularly electric field and ink formulation, droplet sizes can even reach the order of single microns and less—although in the case of bioinks, droplets containing cells will be larger (10s of microns). Using bioinks, cell concentrations on the order of 10^7 cells/mL have been printed with viability after deposition [177], which is an order of magnitude higher than values cited for inkjet- and microvalve-based techniques on relatively accessible components. Pressures used in the process are also low, as the electric field, and not mechanical forces, drives droplet jet formation [169]. Given that droplet diameters can be smaller than the nozzle diameter, the nozzle itself can be bigger to further reduce any pressure required to drive flow of material into the nozzle. High-throughput generation of droplets from a bioink is possible through the continuous nature of this process. However, the major challenge to using EHD droplet generation is the difficulty in achieving high-resolution control of printed structures. Toward this central goal in biofabrication, the continuous, high-throughput generation of droplets makes it difficult to control deposition of any single droplet. The continuous nature of the process that enables high-throughput droplet generation also means that large amounts of material are needed and used, and where recirculation techniques might be used to collect unwanted droplets, there is risk of contamination of the material being printed [239].

LIFT-based bioprinting of bioinks, like EHD and acoustic methods, has strengths in reducing exposure of cells to some of the potentially stressful forces inherent to inkjet and microvalve techniques. It is a nozzle-free technique, where droplets are formed directly from material coated onto a slide, as described earlier. This has the effect of removing nozzle-associated shear forces. LIFT also eliminates the possibility of clogging and allows bioinks with high cell densities, up to ~10^8 cells/mL, to be printed. Ink viscosities up to 300 mPa·s can be used, with high resolution of ~40 μm droplets positioned at a rate of ~5 kHz [185, 240]. High cell survival is reported in printing bioinks [184, 241], although it can be hard to specifically print single cells [240], but this is a challenge that virtually every bioprinting technology faces, and in the case of LIFT methods, computer-vision approaches might allow the targeting of cells on the donor slide for deposition. Like other droplet-based technologies, achieving the highest possible resolution requires bioinks that rapidly gel [241, 242]. Specific to LIFT methods, care must be taken to consider the potential transfer of any material that might have cytotoxicity from the energy absorbing layer into the bioink during droplet formation [243, 244]. Care must be taken to also address technical challenges of achieving a uniform thickness of the bioink coating on the donor slide [184] and to keep it hydrated during the printing process [245], and preparation of this slide may take time if multiple cell types or materials are desired.

1.6.3 Considerations in Energy-Based Bioink Deposition

In the bioprinting technologies classified above as energy-based approaches, the light-based technologies—SLA, DLP, and 2PP—have been used in printing bio-inks. Compared to extrusion- and droplet-based printing of bioinks, the use of bio-inks with these technologies is not currently widespread. As discussed previously, ***SLS*** technology is incompatible with bioinks, as high heat and dry conditions are used in the fabrication process. In the case of the other techniques, SLA is the only technology behind a handful of examples using bioinks before the 2010s [70, 197, 246]. Even in recent years, developing bioinks—or bioresins [87, 198], as a bioink might be referred to in the case of light-based SLA/DLP-type processing—has received significantly less attention than developing bioinks for other bioprinting modalities.

Light-based bioprinting technology development must address shared concerns within bioprinting, for example, developing bioinks with rheological properties suited to the printing technology, which in the case of SLA and DLP approaches should be ideally low-viscosity Newtonian fluids that easily flow under gravity to create a new, homogenous layer of material between each polymerization step and to drain out of the construct from locations where it is not cross-linked [198]. As in other printing modalities, the bioinks should ideally address the challenges of maintaining a homogeneous distribution of cells within the bioink [247, 248], and conditions compatible with the use of bioinks (sterility, minimal print times) should be maintained. However, an additional factor in developing bioinks for light-based methods are many interrelated considerations in applying photoreactive biomaterials simultaneously to a printing process and encapsulating cells [191]. Encapsulating cells within hydrogels using photo cross-linking chemistries is well-established. These encapsulations have been developed using cytocompatible photoinitiators and energy doses. Photoinitiator concentrations, light exposure times, and light intensity can all be controlled to achieve cross-linking of a photo-reactive material without damaging cells [249–252]. While the balance of these parameters is well-established in a static cross-linking environment, in bioprinting this cross-linking must occur in a dynamic layer-by-layer process: over time with one thin layer being cross-linked in direct apposition to the previous layer. Consideration must be given to how energy exposures in one layer may penetrate to the next and also to match energy doses applied in the process to a given bioink. The concentration of the material in the bioink, or bioresin, as well as its chemical formulation, where the concentration and reactivity of the light-reactive functionality, will affect printability—cross-linking, pattern fidelity, rheological properties—in light-based approaches.

In ***SLA-based bioprinting***, the uses of bioinks were demonstrated in the 2000s with high cell viabilities reported [70, 246]. The interwoven considerations of bioink composition and energy application, as mentioned above, must be taken into consideration. In an SLA process, exposure to light at any given location will depend on the scanning/rastering speed of the laser and its intensity. There exists potential for light to penetrate to previously exposed layers, which can benefit layer-to-layer adherence, as discussed earlier, but should be considered with respect to energy dosages applied to a bioink. Macroscale constructs (>1 cm^3) with feature resolution as high as 20 μm are possible using this technology, offering fantastic multiscale potential. Realizing this potential and moving toward more complexity in printed tissue constructs will require further bioink and process development. Achieving discrete high-resolution elements within a single layer requires that the process allows for exchanging bioinks while printing a single layer at a given z-position, a challenge that would scale with increasing resolution. While typical commercial SLA hardware uses a single material for an entire construct, a couple of examples of the potential for this process modification examples exist in closely related DLP-based bioprinting [253, 254], and also using SLA bioprinting where bioinks were varied from one layer to the next [247].

Considerations and challenges using ***DLP-based bioprinting*** methods are generally the same, as in SLA [24]. The primary difference between the technologies is how light is applied, with DLP offering the potential to cure entire layers simultaneously and thus speed up print times considerably. There are potential downsides in terms of the quality of light delivered to a surface depending on the DLP technology used—as discussed earlier—and SLA has been shown to offer the potential to alter the laser energy during the printing process to achieve different results on a point-by-point basis [246]. However, DLP technology combines speed with high-resolution capabilities that have been used to print features in the 25–50 μm range in work developing bioinks [255] for DLP bioprinting [198].

DLP-based methods have been used in demonstrating the potential of longer wavelength (visible) light with bioinks in bioprinting [200, 256], where additives were also included in the bioink to match the buoyant density of the cells to prevent sedimentation [200], similar to approaches used in inkjet printing [171]. Owing to the similarity to laser-scanning SLA approaches, advances addressing challenges associated with one technology should generally translate easily to the other. As mentioned above, DLP-based methods have also been used to print multiple bioinks, on the same layer, to achieve biomimetic cellular and material complexity at high resolution [253, 254]. In the more recent example, custom microfluidic hardware—in which a flexible membrane served as a build platform that could be moved, step-wise, in z-space—offered the capabilities to rapidly change bioinks to achieve this complex heterogeneity [254]. While on a small scale, this sort of system may have strengths in terms of minimizing material waste and controlling the environment to which cell-containing bioinks are exposed.

In an earlier microfluidic approach, an approach similar to SLA-based bioprinting was used to cross-link multimaterial and cell-containing hydrogel structures at a microscale resolution [257]. However, this approach used a confocal laser to scan within a space, and establishing material structure extending in the z-direction was limited by the confocal hardware's depth of focus. Similarly, ***two-photon polymerization***, as a bioprinting technique, has also been limited to use with very small structures, and cells, when included in the final construct, are often added after fabrication. The limitations in 2PP on being able to fabricate structures whose size exceeds the millimeter size range impact its usefulness as a platform for engineering larger tissue for implantation, despite resolution capabilities that allow the fabrication of microenvironmental features that can influence cells at a single-cell scale for in vitro assays [211]. Nonetheless, as discussed earlier, 2PP has unparalleled capabilities for printing structures at high resolution, and recent work has demonstrated this promise in the printing of bioinks, enabling cells to be printed at high resolution with high viability [258]. Work toward improving the bioink for use with 2PP has included the development of photoinitiators [259] and the cross-linking chemistry [260]. The development of bioinks can be challenging for this technique, as the adjustment of hardware parameters to different materials can be time consuming [261].

1.7 Expanding the Capabilities of Bioprinting Through Innovation and Closely Related Technologies

Bioprinting technologies are enabling the fabrication of engineered biological and tissue structures with increasing complexity and biomimetic function. As discussed in the previous section, each bioprinting modality has strengths and potential limitations. Further development of bioprinting technologies and techniques—which are often enabled by the development of materials for bioinks and through applications of bioprinting that yield fundamental insights into engineering tissue—will continue advance capabilities in tissue engineering and regenerative medicine. Specific materials, as well as applications of bioprinting, will be discussed later in the book. Some challenges within bioprinting are shared across multiple printing modalities, including the need for bioinks that have the rheological properties required for SFF, the technical challenges inherent to layer-by-layer printing, and the limitations in adapting additive manufacturing technologies that were generally designed for acellular materials to the printing of highly hydrated bioinks. Other challenges are not unique to bioprinting but are shared broadly across tissue engineering technologies, and these include the creation of channels—or vascular structures—within larger constructs, the need to achieve biomimetic structures and function, and challenges associated with reaching cell densities of native tissues. Ultimately, these challenges may all have to be solved simultaneously to realize bioprinted tissues and organs.

1.7.1 Multimodal and Multimaterial Printing

Using multiple bioprinting modalities and material types in the fabrication of a tissue construct can combine the strengths of multiple technologies to address limitations inherent to a single method. For example, the lack of stability in a highly hydrated bioink can be addressed by first printing a more mechanically robust material that can define and maintain the bulk shape of the construct. This printed support material can then, in spaces between its filaments, for example, hold one or more bioinks in place, as has been demonstrated by first extruding a thermoplastic and then soft bioinks, which are extruded between the thermoplastic filaments, one layer after another [1, 214, 262] (Fig. 1.10b). This approach has been demonstrated with multiple soft materials combined in the same construct, where no mixing of the soft hydrogels was observed and where cells within deposited bioinks were observed to be unharmed by heat from the thermoplastic filaments deposited in the next layers [214]. A thermoplastic component can also offer increased mechanical properties where a higher compressive modulus might be desired for a cell-containing construct than a cell-containing bioink can provide [263].

Materials used in printing filaments, between which bioinks might be deposited are not limited to thermoplastics, but can also include soft materials that can be directly written by extrusion, but which might not contain cells themselves. A softer material may be desirable as a printed support in bioprinting flexible soft tissue constructs and in controlling degradation [264]. Not all bioinks lack stability upon deposition, and thus a softer support material can also be a cell-containing bioink that rapidly stabilizes. This has been demonstrated in a multi-modal bioprinting approach that combined a bioink support matrix with the placement of cell spheroids into the spaces between filaments [265]. This addresses one challenge facing cell spheroid-based bioprinting—the need for a support structure or material during printing and while the spheroids mature and merge. The material structure potentially offers additional control over the shape and maturation of a spheroid-based construct. Similar work has developed complex scaffolds by combining thermoplastic extrusion of a biodegradable material with the placement of cell spheroids, enabled in part through the design of microfluidic controls capable of handling cell spheroids and delivering one at a time [215] (Fig. 1.10c).

In other examples of how the strengths of multiple technologies can be combined, FDM, bioink extrusion, and inkjet printing of a bioink have been used to create spatial organized tissue constructs [266]. Thermoplastic material provided support for bioinks and cells, as well as desirable mechanical properties for the bulk construct. Microextrusion and inkjet printing each allowed the deposition of cells, with each modality depositing bioinks in separate regions of the material and inkjetting being used to tightly control the number of cell regions where cells would ultimately mature into spheroids with cartilage-like properties. Other work has combined extruded thermoplastic and SLA-printed bioinks to create a rigid support around a vessel-like structure, containing living vascular cells, that could be further surrounded with hydrogel material in a TE construct [267].

While all these examples have all included extrusion-based printing to fabricate a support structure, multimodal approaches to bioprinting tissue constructs can leverage many bioprinting and biofabrication modalities. A hybrid printing system has been demonstrated combining printheads capable of inkjet printing and electrospinning—a processing method similar to EHD droplet deposition that deposits fibers with diameters on the nanoscale homogeneously across a surface [268]. Inkjet printing offered controlled delivery of cells and biological hydrogel support materials, and electrospun fibers between each inkjet printed layer provided stability and support for the bioink. In another approach, LIFT bioprinting has similarly been used to deposit bioinks onto scaffolds fabricated to contain microscale features using 2PP techniques [269]. In these cases, electrospun fibers or 2PP printed scaffolding provide support and topographic cues at scales that exceed capabilities in standard extrusion techniques.

Melt electrowriting [84, 85], a relatively young technology, blends strengths of extrusion-based bioprinting and electrospinning. It offers extrusion's 3D positioning capabilities and but with micron-diameter filaments whose scale approaches that achievable in electrospinning. In this technique, heat and charge are applied to a nozzle to allow a polymer melt to leave a nozzle and be stably collected in a layer-by-layer fashion. While this technique is strictly acellular, it is closely akin to 3D bioprinting and is representative of technological innovation that can continue to push SFF capabilities. Scaffolds created using MEW have been used to help control the morphology of vascular networks [270] and to increase the mechanical strength of cell-containing hydrogel materials [271, 272].

1.7.2 Embedded Bioprinting and Addressing Challenges Inherent to Layer-by-Layer Printing

A recurring challenge addressed in the innovative bioprinting approaches discussed above is supporting bioinks upon deposition. Bioinks are generally soft, highly hydrated materials with relatively low polymer content, and the rate at which many of the materials solidify upon leaving a print nozzle is slower than the rate at which they flow away from the point of deposition. One emerging technique to enable the printing of diverse materials, including bioinks and cell suspensions, is the printing of ink materials directly into a permissive support material in a method referred to as embedded printing or printing within a suspension medium. In embedded printing, the volume into which the ink material will be printed is first filled with a material which is capable of rearranging to allow a print nozzle to move through it. This support material should also be designed to recover, or restabilize, after the print nozzle has passed through a particular location, holding any deposited ink—or bioink—in place (Fig. 1.11). Embedded approaches can be used in bioprinting arbitrary structures of great complexity—with large overhangs, internal void spaces, and nested structures (objects within objects) [112, 113, 126, 274, 275]. By using a support medium, these structures can be directly printed, with the extruded ink held

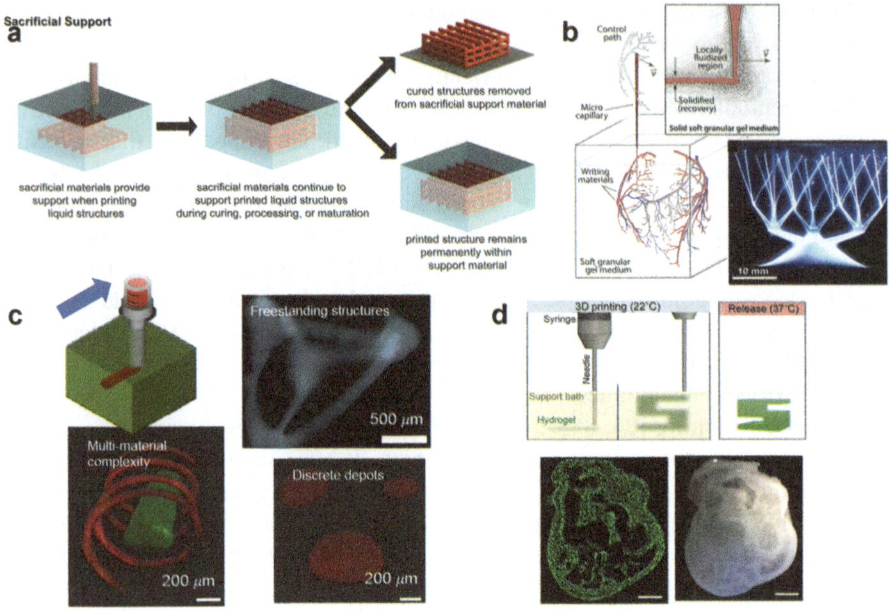

Fig. 1.11 (**a**) Printing bioinks into support materials that might be sacrificial or a permanent part of the final construct. (**b**) Multiscale structures printed into jammed, granular medium. (**c**) Printing of a hydrogel into a hydrogel. (**d**) Printing of heart construct with internal voids into particle-based support that can be melted away at physiological temperatures ((**a**) adapted from O'Bryan et al. MRS Bull 2017 [273], (**b**) adapted from Bhattacharjee et al. Sci Adv 2015 [112], (**c**) adapted from Highley et al. Adv Mater 2015 [126], (**d**) adapted from Hinton et al. Sci Adv 2015 [113])

in place during gelation, allowing diverse ink formulations to be used [112]. This offers advantages over traditional bioprinting approaches, where complex structures with overhanging features can be printed, but the process must include the layer-by-layer deposition of a removable support material, and challenges surrounding bio-ink solidification would remain [114].

Printing within suspension media can support the printing of a construct that is subsequently removed from the support material, as described, but it can also be used in printing where the support material remains as part of the final construct. In these cases, the suspension medium provides ongoing support, after printing, to printed features that might include materials [126], cells [112, 276], or to void spaces that might be lined with cells to form vascular structures within a construct [118, 126, 277]. Suspension media have included materials that recover through the introduction of new material into a defect the needle creates [277] followed by a reaction to cure the defect. Other support materials self-heal once the print nozzle has moved away from a particular location within the support [112, 113, 126].

Embedded printing methods offer opportunities to address challenges both specific to bioprinting processes as well as to TE more broadly. These techniques are not limited to printing one layer at a time. Any print location within the support material is accessible for bioink deposition. Continuous material features can easily be printed in any direction through 3D space, instead of as discrete elements, in traditional layer-by-layer approaches. Bioprinting occurs within a hydrated environment. Through the use of support media that are designed to be a part of the final construct, not ultimately removed, embedded printing may also address challenges of print times and vascularization. A support medium which remains might include homogeneously distributed cells, as in standard tissue engineering approaches, but be viewed as a 3D canvas, where bioprinting might be used to deposit only the high-resolution features necessary to create the desired heterogeneity within a tissue construct. In standard layer-by-layer approaches, print times increase exponentially with increasing resolution [224, 273], putting practical limits on feature resolution, which is often much less than the positioning capabilities of bioprinting hardware.

This ability to use a support to print heterogeneity as needed, at high resolution, is coupled with the ability to print vasculature-like void spaces that can branch and rejoin [118, 126, 277]. This requires an ink material which can be printed into the support and subsequently removed—a sacrificial [273], or fugitive [40], ink—and a support material that is designed resist collapse after the removal of the fugitive material. Printing into suspension media thus offers strengths both in fabrication of structures that might be removed from the media and also in the design of constructs that include these media. In the latter case, the printing process is able to harness high-resolution capabilities and establish critical channels for nutrient exchange in the final construct.

Addressing the grand challenge of printing void spaces or vasculature within a TE construct is thus possible through embedded methods as well as by leaving void spaces within printed constructs using standard bioprinting processes. A related approach is the use of sacrificial inks, deposited by bioprinters in vessel-like conformations, around which a hydrogel is then cured, which is followed by the removal of the sacrificial structure, leaving void spaces within the hydrogel (Fig. 1.12). This leverages printer capabilities to create multiscale filament structures within diverse materials [124]. These structures might be printed to have complex, 3D geometries and can be lined with endothelial cells to create models of vasculature [125, 126, 279, 280]. In many of these approaches, printing is used to establish the geometry of a sacrificial material around which a homogeneous cell-containing hydrogel might be cured. But the capabilities of bioprinters have also been harnessed in the printing of heterogeneous, multicellular structures around a sacrificial ink, prior to surrounding printed structures with a hydrogel [278], demonstrating the potential to combine fugitive inks with mulitple bioinks in these methods to harness the strengths of bioprinting in creating vascularized macroscale TE constructs [125].

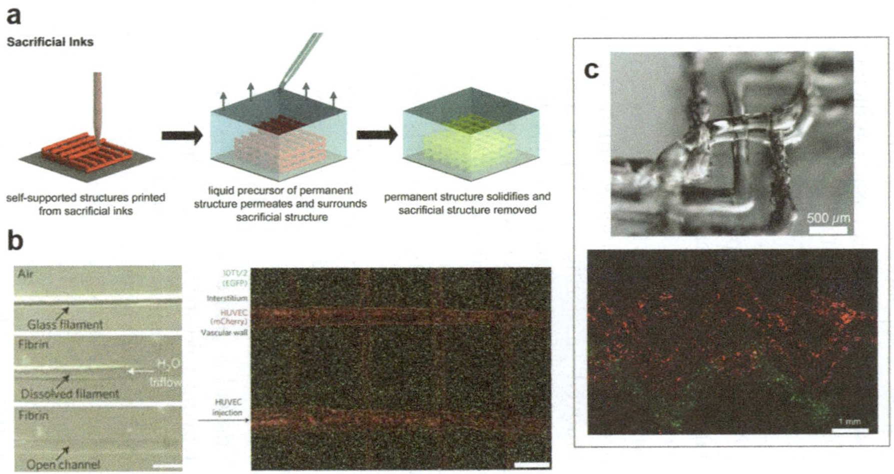

Fig. 1.12 (**a**) Printing removable bioinks that are embedded within support material and subsequently removed to leave channels for nutrient exchange. (**b**) Printing of sugar filaments, lining of channels with endothelial cells. (**c**) Printing of a removable filament surrounded by bioinks containing multiple cell types, followed by embedding in a hydrogel and removal of the sacrificial filament ((**a**) adapted from O'Bryan et al. MRS Bull 2017 [273], (**b**) adapted from Miller et al. Nat Mater 2012 [124], (**c**) adapted from Kolesky et al. Adv Mater 2014 [278])

1.7.3 Modifying the Printing Process and Hardware to Extending Bioprinting Capabilities and Materials

Innovation driving bioprinting capabilities is often related both the hardware and bioinks used. Steering clear of specifics for materials, which will be discussed at length in the following chapter, innovation in nozzle design is a particular example of how hardware innovation can enhance bioprinting capabilities and possibilities. Where the techniques discussed above—multimodal printing and embedded printing—often address challenges related to bioink behaviors after leaving a printer's nozzle, modifications to the nozzle itself can improve printer capabilities and additional opportunities for controlling multiscale features within printed constructs. One example of this is the use of coaxial, or concentric, nozzles that can be used to deliver multiple materials.

Coaxial nozzles can be used to print hollow filaments that are useful in establishing channels for nutrient exchange or patterning vascular structures. These nozzles can deliver multiple materials—usually two—in a core-shell arrangement [281], so that the resulting filament results from the interactions of these two materials. By delivering a cross-linkable bioink through the external nozzle and the cross-linking agent through the internal nozzle, hollow filaments can be bioprinted [218, 265, 282–285] (Fig. 1.13a). Alternatively, a cross-linking agent might be delivered in the shell and initiate gelation of the core material, resulting in a solid filament [223]. Bioinks can be designed to deliver cell types specific to vasculature in these approaches, as will be discussed at greater length in later chapters. Coaxial nozzles also offer capabilities for depositing multiple materials, but one at a time, to control material distribution at the meso- and microscale along a printed filament [218]. Such capabilities are perhaps best demonstrated using microfluidic devices and control systems that can precisely combine and deliver multiple input inks to a nozzle [223, 286, 288].

Many of the innovations in nozzle design have been focused on improving extrusion-based processes. In these bioprinting processes especially, the nozzle represents the location in the process where a bioink should ideally transition from a flowing, liquid-like state to a gelled, solid-like state. The design of nozzles in which bioink gelation might be initiated, and where flow rates are matched to gelation times, might allow the bioprinting of materials with slower gelation rates. Bioinks that gel slowly (on the order of seconds to minutes) are often not compatible with the print processes, because the stabilization of material must occur almost instantaneously upon leaving the nozzle. Photopermeable nozzles have been used to allow the direct extrusion of bioinks which are cross-linked in the presence of light [218] (Fig. 1.13b). There may exist opportunity for nozzles which can initiate other forms of cross-linking as materials move through them. Heated nozzles and print platforms are common, but hardware supportive of enzymatic and chemical reactions that occur on longer timescales might be considered. A nozzle capable of combining materials with active mixing has been used to control heterogeneity in material composition during continuous extrusion [289], and similar technology might be

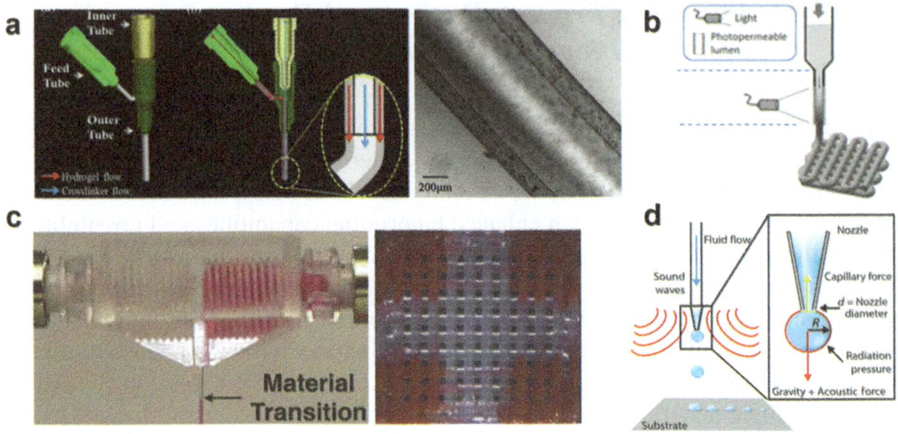

Fig. 1.13 Innovation in hardware design at print nozzles. (**a**) Coaxial nozzle (left) with hollow, cell-containing filament (right). (**b**) Photopermeable nozzle for photo cross-linking during extrusion. (**c**) Microfluidic nozzle where inks can meet and therefore be continuously altered during printing to create hetereogeniety. (**d**) Acoustophoretic printing, allowing a very broad range of material viscosities to be printed ((**a**) adapted from Zhang et al. Biofabrication 2013 [282] and Gao et al. Biomaterials 2015 [284], (**b**) adapted from Ouyang et al. Adv Mater 2017 [218], (**c**) adapted from Hardin et al. Adv Mater 2015 [286], (**d**) adapted from Foresti et al. Sci Adv 2018 [287])

advantageous in combining bioink components that form gels through chemical reactions (Fig. 1.13c).

It is not only extrusion-based processes where hardware innovation can improve processing by bioprinting. Although examples are more limited in droplet- and energy-based approaches, continued advances are pushing bioprinting and biofabrication forward by addressing weaknesses that hinder broader application. For example, a major challenge in droplet-based bioprinting is the requirement that low viscosity bioinks are used. By designing hardware capable of generating accurate, localized acoustophoretic forces in air, uniform droplets of liquids, including bioinks with viscosities ranging from 0.5 to 25,000 mPa·s have been ejected off the end of a nozzle [287] (Fig. 1.13d). In an adaptation to droplet-based printing that is in some ways reminiscent of bioplotting, droplets have been formed within liquid environments and patterned to form tissue-like structures. This work included bioinks and yielded well-defined 3D structures and tissue constructs from soft materials, without the need for rapid cross-linking upon reaching a print surface [290, 291].

1.7.4 Technologies at the Boundary Between Bioprinting and Biofabrication

Delineating what technologies fit within the realm of bioprinting and which belong to perhaps another class of biofabrication technologies is a problem that may await the development of broadly established standards. In the meantime, just as classifications of bioprinting modalities vary among sources, technologies considered to be bioprinting likely will as well. As the field continues to innovate, capabilities will inevitably cross boundaries, and it seems likely that continued work combining existing technology, in addition to building new, will be essential to advances.

Two-photon polymerization bioprinting technologies, for example, are closely related to two photon-based techniques used with photodegradable hydrogels. Just as 2PP can be used to cross-link structures, as discussed, at very high resolution, given a material that is cross-linked through photolabile chemistries, two-photon light can be used to degrade material in a selective way [203, 204] (Fig. 1.14a). Although quite the opposite of additive manufacturing, this subtractive technique has been used to establish high-resolution channels in cell-containing hydrogels for the purposes of influence cell behaviors [203]. In using laser-based approaches to creating channels, photodegradable chemistries are not even required within some hydrogels—nano- and femtosecond pulsed lasers can directly ablate a hydrogel structure [205]. However, like other lithographic technologies, two photon-based lithography

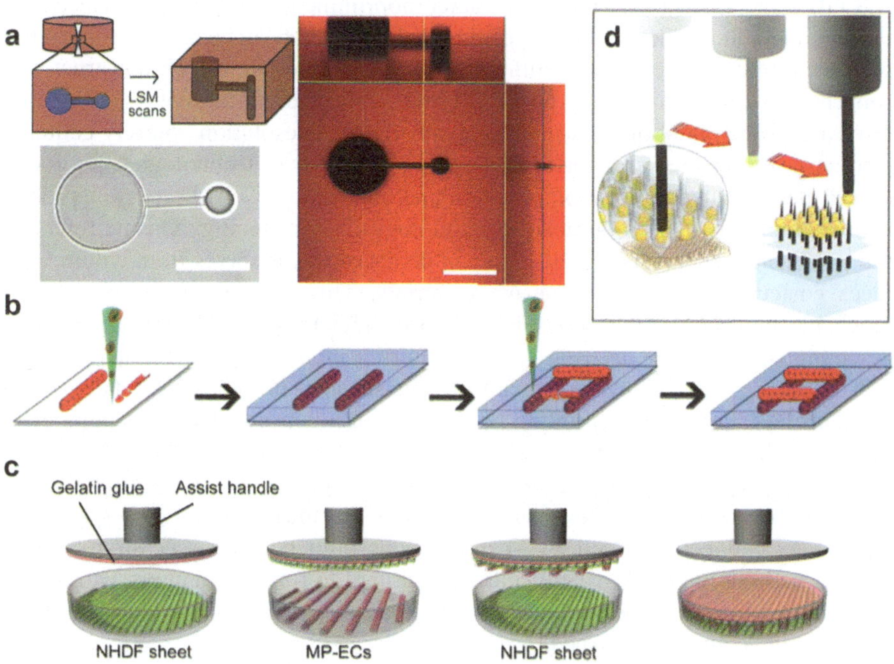

Fig. 1.14 Biofabrication and bioassembly technologies that are very closely related to bioprinting technologies. (**a**) Two-photon patterning of channels in hydrogels. (**b**) Laser-guided direct writing—a bioprinting technology. (**c**) Multi-layer constructs fabricated from cell sheets. (**d**) Microneedles supporting the scaffold-free assembly of tissue constructs from cell spheroids ((**a**) adapted from Kloxin et al. Science 2009 [203], (**b**) adapted from Nahmias et al. Biotechnol Bioeng 2005 [292], (**c**) adapted from Tsuda et al. Biomaterials 2007 [293], (**d**) adapted from Ong et al. Sci Rep 2017 [294])

can also be used to conjugate biofunctionalities to a hydrogel network at high resolution through photoreactive chemistries [206]. This comes closer to an additive manufacturing technique, though the cell-containing material is already cross-linked and light is used to further modify it, rather than to solidify and print it from a flowing ink. In short, in light of 2PP being a powerful technique for bioprinting microscale structures, it also bears noting that through materials design, this technology might be used in many other potentially synergistic ways in a biofabrication process.

Laser-guided direct writing (LGDW) is an example of a technology that might be considered bioprinting, but at a minimum is very closely related and which has unique and intriguing capabilities for structuring tissue constructs [38]. Like many bioprinting modalities, LGDW can directly manipulate and position small biological materials, including individual cells, on surfaces. The technology uses an optical force that arises when an object with a high refractive index is within the light beam. The object becomes radially trapped within the beam and will move axially along it [292] (Fig. 1.14b). As in other cell-based approaches (e.g. spheroids), support must be provided, often using a material coating, to facilitate the stable placement of cells on a surface and in subsequent layers, as cells are positioned upon one another in a construct.

LGDW therefore joins a number of other bioprinting technologies described previously, including spheroid-based printing and LIFT, that can be used to control the placement of cells, ideally with minimal additional material, in a TE construct—in some cases with resolution approaching a single cell. These technologies offer paths toward a major biofabrication goal of achieving high-resolution, densely cellularized tissue-engineered structures. It is among a number of technologies that focus on cell-dense materials for biofabrication and bioprinting. Cell sheets are a prominent example of such biofabrication technology. The technology's name refers to the confluent monolayers of cells that can be harvested as sheets from culture surfaces coated with thermoresponsive materials. This technology has developed alongside bioprinting and cell spheroid-based technologies and can be used to assemble structures by layering sheets, requiring minimal support material [293, 295, 296] (Fig. 1.14c). Layered assembly using sheets is in some ways analogous to standard layer-by-layer bioprinting techniques. In a convergence of technologies, cell sheets have been harvested and processed for using in bioprinting [297], with the sheets being broken into small fragments that serve as the basis for a bioink in extrusion-based printing. The resulting bioinks are thus cell-rich like spheroids, with many cell–cell and cell–ECM interactions influencing. Tissue strands offer another cell-dense bioink, where cells are cultured to form continuous, extrudable cylinders—as opposed to discrete spheroids—composed of cells that may self-assemble into functional tissue [145].

At the level of individual cells, cell surface modifications have been shown to facilitate the assembly of larger tissue structures and may have application in conjunction with bioprinting technologies discussed above. Thin hydrogel coatings on cells have resulted in thin, adherent layers of cells, reminiscent of cell sheets, in which microvascular networks form based on cell composition [298]. By using patterns of single-stranded DNA patterned on a substrate, cells whose surfaces are modified with complementary single-stranded DNA can be assembled with great precision [299] (<10 μm resolution). These are cell-dominant fabrication technologies, like cell sheets and spheroids. Although these cell-based assembly approaches may find application in connection with bioprinting, they constitute unique biofabrication technologies than aim to build a tissue from basic components, from the bottom up. There are many technologies within biofabrication that fall within this category, and covering them is beyond the scope of the chapter, but the interested reader is encouraged to refer to broader reviews of biofabrication [6, 54, 91].

Other technologies that sit at the boundary of those that would be considered bioprinting include robotic technologies that are capable of picking up and then placing individual microtissue units onto a construct on a build platform. Like spheroids and tissue strands, microtissues are derived from cells cultured under conditions that cause them to aggregate into units of fabrication—in this case, through culture in toroidal- or honeycomb-shaped molds [300]. Similarly, cell spheroids have been bioprinted onto tight arrays of microneedles (Fig. 1.14d). Obviating the need for support material, these microneedles are closely spaced such that spheroids placed onto them are in contact with adjacent spheroids to enable interactions and the maturation of the bulk tissue structure [148, 294]. The variety of ways in which bioprinting and related technologies might be used will continue to multiply, and it seems clear that synergies between technologies—for example, even between the apparently mutually exclusive scaffold-free and scaffold-based approaches [301]—will be important to advances.

1.8 Looking Toward the Future of Bioprinting

Bioprinting is currently in an exciting and active stage of technological development. There exist numerous challenges within bioprinting itself as well as within broader fields related to tissue engineering. But, as has been discussed here, there are also substantial opportunities to innovate and address these challenges. With respect to tissue engineering, bioprinting offers a combination of strengths that are well suited to addressing challenges of engineering complexity and function in tissue constructs—both now and through continued advances. In the following chapters, examples of the current capabilities and potential of the bioprinting technologies introduced here will be covered through areas of active application of bioprinting technology. These chapters will illustrate how the technologies described in this chapter are currently allowing us to move beyond simplistic 2D in vitro model systems, which are often not able to recapitulate the environments needed to answer critical questions within TE [302] toward the controlled fabrication of not just homogeneous 3D gels, but complex 3D constructs. Leveraging the technology's capabilities to engineer-controlled biological complexity has been a central goal of bioprinting since the earliest efforts to apply printer technology to biological questions [16].

Currently, and in the near future, bioprinting will likely have its greatest translational impact through enabling the fabrication of in vitro models that recapitulate native biology or physiology. From in vitro models and, for example, "on a chip"-type devices, the importance of using fabrication techniques to establish biomimetic conditions and cellular arrangements is well appreciated [303]. Bioprinted in vitro systems are already enabling drug screening on engineered tissues that mimic human responses to drugs [304]. Bioprinting technologies will also continue to interact synergistically with other fields. From stem cell biology, for example, we know that cell fates depend heavily on microenvironmental cues [305], and fabrication technologies that allow us to define these environments [115] will help researchers ask and answer fundamental questions within their fields. In turn, the results of such work will help us understand what we must be capable of bioprinting and how we might provide the appropriate cues to stem cells to guide their development [44] into functional structures—one possible route to engineering replacement tissue.

Closely related to both stem cells and cell spheroids are organoids, 3D aggregates of pluripotent stem cells that, with the correct cues, can give rise to primitive organ-like structures [12]. The potential for bioprinting or biofabrication technologies to accelerate organoid research and for organoid research to accelerate the fabrication of functional tissue seems high. Challenges faced in the maturation of organoids beyond current limits include the need for more complex architectures and cellular compositions and the need for vasculature [4, 306]. These requirements echo those of tissue engineering and goals of researchers working in bioprinting and biofabrication. Bioprinting technologies may offer the capabilities needed to create constructs that guide the self-assembly of tissue from organoids. Bioprinting may also be used to directly control the placement of primitive tissue and material structures in ways that might facilitate their own inherent abilities to give rise to increasingly complex tissue structures [307, 308].

In the clinic, in the nearer term, bioprinting might offer a technology for delivering cell-based therapies and cell-containing biomaterials directly to an injury site. In situ bioprinting approaches have directly printed bioinks into skin wounds in animals [309, 310]. These capabilities might be extended to deliver complex, structured regenerative materials into patient-specific wound sites in the clinic, as an alternative to prefabricating a construct or injecting a material without control over the features of that material within the volume of the injury site. Printing functional tissues and whole organs remains the long-term goal that motivates many researchers in the field. Perhaps through breakthroughs in combinations of bioprinter innovation, advances in basic science (e.g., stem cell biology and developmental biology), and capabilities in replicating critical microevironmental cues at the appropriate resolutions, achieving this goal may be possible sooner rather than later. Innovation is unpredictable. Of course, we will have to remain cognizant of ethical concerns that might follow from recreating human structures—particularly neural structures. Challenges will also surround cell sourcing—the expansion and culture of sufficient cell numbers to recreate tissue is a major consideration [311]. Finally, federal regulations surrounding bioprinted products that both safeguard patients and provide clear frameworks for the development of bioprinted products will be important [312].

How bioprinting will be used—how large of a role it will ultimately play in biomedicine—remains to be seen. The impacts of the technology are being felt now. Bioprinting is advancing, and the possibilities still seem limited only by the extent of our knowledge and our ingenuity.

References

1. Kang H-W, Lee SJ, Ko IK et al (2016) A 3D bioprinting system to produce human-scale tissue constructs with structural integrity. Nat Biotechnol 34:312–319
2. Ingber DE, Mow VC, Butler D et al (2006) Tissue engineering and developmental biology: going biomimetic. Tissue Eng 12:3265–3283. https://doi.org/10.1089/ten.2006.12.3265
3. Lenas P, Moos M, Luyten FP (2009) Developmental engineering: a new paradigm for the design and manufacturing of cell-based products. Part I: from three-dimensional cell growth to biomimetics of in vivo development. Tissue Eng B Rev 15:381–394. https://doi.org/10.1089/ten.teb.2008.0575
4. Yin X, Mead BE, Safaee H et al (2016) Engineering stem cell organoids. Cell Stem Cell 18:25–38. https://doi.org/10.1016/j.stem.2015.12.005
5. Groll J, Boland T, Blunk T et al (2016) Biofabrication: reappraising the definition of an evolving field. Biofabrication 8:013001. https://doi.org/10.1088/1758-5090/8/1/013001
6. Moroni L, Burdick JA, Highley C et al (2018) Biofabrication strategies for 3D in vitro models and regenerative medicine. Nat Rev Mater 3:21–37. https://doi.org/10.1038/s41578-018-0006-y
7. Viola J, Lal B, Grad O (2003) NSF: Abt report on "The Emergence of Tissue Engineering as a Research Field". https://www.nsf.gov/pubs/2004/nsf0450/start.htm. Accessed 23 Feb 2019
8. Su A, Al'Aref SJ (2018) Chapter 1—History of 3D printing. In: Al'Aref SJ, Mosadegh B, Dunham S, Min JK (eds) 3D printing applications in cardiovascular medicine. Academic, Boston, pp 1–10
9. Langer R, Vacanti JP (1993) Tissue engineering. Science 260:920–926. https://doi.org/10.1126/science.8493529
10. Elliott NT, Yuan F (2011) A review of three-dimensional in vitro tissue models for drug discovery and transport studies. J Pharm Sci 100:59–74. https://doi.org/10.1002/jps.22257
11. Tam RY, Yockell-Lelièvre J, Smith LJ et al (2019) Rationally designed 3D hydrogels model invasive lung diseases enabling high-content drug screening. Adv Mater 31:1806214. https://doi.org/10.1002/adma.201806214
12. Lancaster MA, Knoblich JA (2014) Organogenesis in a dish: modeling development and disease using organoid technologies. Science 345:1247125. https://doi.org/10.1126/science.1247125
13. Place ES, Evans ND, Stevens MM (2009) Complexity in biomaterials for tissue engineering. Nat Mater 8:457–470. https://doi.org/10.1038/nmat2441
14. Mao AS, Mooney DJ (2015) Regenerative medicine: current therapies and future directions. Proc Natl Acad Sci 112:14452–14459. https://doi.org/10.1073/pnas.1508520112
15. Hunsberger J, Harrysson O, Shirwaiker R et al (2015) Manufacturing road map for tissue engineering and regenerative medicine technologies. Stem Cells Transl Med 4:130–135. https://doi.org/10.5966/sctm.2014-0254
16. Klebe RJ (1988) Cytoscribing: a method for micropositioning cells and the construction of two- and three-dimensional synthetic tissues. Exp Cell Res 179:362–373

17. Klebe RJ, Thomas CA, Grant GM et al (1994) Cytoscription: computer controlled micropositioning of cell adhesion proteins and cells. J Tissue Cult Methods 16:189–192. https://doi.org/10.1007/BF01540648
18. Cesarano J, Baer TA, Calvert P (1997) Recent developments in freeform fabrication of dense ceramics from slurry deposition. International Solid Freeform Fabrication Symposium, 25–32.
19. Cesarano J, Segalman R, Calvert P (1998) Robocasting provides moldless fabrication form slurry deposition. Ceram Ind 148:94
20. Maruo S, Nakamura O, Kawata S (1997) Three-dimensional microfabrication with two-photon-absorbed photopolymerization. Opt Lett 22:132–134. https://doi.org/10.1364/OL.22.000132
21. Kawata S, Sun H-B, Tanaka T, Takada K (2001) Finer features for functional microdevices. Nature 412:697–698. https://doi.org/10.1038/35089130
22. Ma PX, Elisseeff J (2005) Scaffolding in tissue engineering. CRC Press, Boca Raton
23. Lewis JA (2006) Direct ink writing of 3D functional materials. Adv Funct Mater 16:2193–2204. https://doi.org/10.1002/adfm.200600434
24. Billiet T, Vandenhaute M, Schelfhout J et al (2012) A review of trends and limitations in hydrogel-rapid prototyping for tissue engineering. Biomaterials 33:6020–6041. https://doi.org/10.1016/j.biomaterials.2012.04.050
25. Smay JE, Cesarano J, Lewis JA (2002) Colloidal inks for directed assembly of 3-D periodic structures. Langmuir 18:5429–5437. https://doi.org/10.1021/la0257135
26. Chrisey DB (2000) The power of direct writing. Science 289:879–881. https://doi.org/10.1126/science.289.5481.879
27. Lewis JA, Gratson GM (2004) Direct writing in three dimensions. Mater Today 7:32–39. https://doi.org/10.1016/S1369-7021(04)00344-X
28. Wikipedia (2019) Robocasting. https://en.wikipedia.org/wiki/Robocasting
29. Griffith LG, Naughton G (2002) Tissue engineering—current challenges and expanding opportunities. Science 295:1009–1014. https://doi.org/10.1126/science.1069210
30. Wu BM, Borland SW, Giordano RA et al (1996) Solid free-form fabrication of drug delivery devices. J Control Release 40:77–87. https://doi.org/10.1016/0168-3659(95)00173-5
31. Giordano RA, Wu BM, Borland SW et al (1996) Mechanical properties of dense polylactic acid structures fabricated by three dimensional printing. J Biomater Sci Polym Ed 8:63–75
32. Park A, Wu B, Griffith LG (1998) Integration of surface modification and 3D fabrication techniques to prepare patterned poly(L-lactide) substrates allowing regionally selective cell adhesion. J Biomater Sci Polym Ed 9:89–110. https://doi.org/10.1163/156856298X00451
33. Lee G, Barlow JW (1993) Selective laser sintering of bioceramic materials for implants. International Solid Freeform Fabrication Symposium, 376-380.
34. Hutmacher DW, Schantz T, Zein I et al (2001) Mechanical properties and cell cultural response of polycaprolactone scaffolds designed and fabricated via fused deposition modeling. J Biomed Mater Res 55:203–216. https://doi.org/10.1002/1097-4636(200105)55:2<203::AID-JBM1007>3.0.CO;2-7
35. Bhatia SN, Chen CS (1999) Tissue engineering at the micro-scale. Biomed Microdevices 2:131–144. https://doi.org/10.1023/A:1009949704750
36. Bhatia SN, Balis UJ, Yarmush ML, Toner M (1998) Probing heterotypic cell interactions: hepatocyte function in microfabricated co-cultures. J Biomater Sci Polym Ed 9:1137–1160. https://doi.org/10.1163/156856298X00695
37. Chen CS, Mrksich M, Huang S et al (1997) Geometric control of cell life and death. Science 276:1425–1428. https://doi.org/10.1126/science.276.5317.1425
38. Odde DJ, Renn MJ (1999) Laser-guided direct writing for applications in biotechnology. Trends Biotechnol 17:385–389. https://doi.org/10.1016/S0167-7799(99)01355-4
39. Wilson WC, Boland T (2003) Cell and organ printing 1: protein and cell printers. Anat Rec A Discov Mol Cell Evol Biol 272A:491–496. https://doi.org/10.1002/ar.a.10057

40. Therriault D, White SR, Lewis JA (2003) Chaotic mixing in three-dimensional microvascular networks fabricated by direct-write assembly. Nat Mater 2:265–271. https://doi.org/10.1038/nmat863
41. All3DP (2016) The official history of the RepRap project. In: All3DP. https://all3dp.com/history-of-the-reprap-project/. Accessed 23 Feb 2019
42. Malone E, Lipson H (2007) Fab@Home: the personal desktop fabricator kit. Rapid Prototyp J 13:245–255. https://doi.org/10.1108/13552540710776197s
43. Feinberg AW, Miller JS (2017) Progress in three-dimensional bioprinting. MRS Bull 42:557–562. https://doi.org/10.1557/mrs.2017.166
44. Lutolf M, Hubbell J (2005) Synthetic biomaterials as instructive extracellular microenvironments for morphogenesis in tissue engineering. Nat Biotechnol 23:47
45. Engler AJ, Sen S, Sweeney HL, Discher DE (2006) Matrix elasticity directs stem cell lineage specification. Cell 126:677–689. https://doi.org/10.1016/j.cell.2006.06.044
46. Li L, Eyckmans J, Chen CS (2017) Designer biomaterials for mechanobiology. Nat Mater 16:1164–1168. https://doi.org/10.1038/nmat5049
47. Burdick JA, Murphy WL (2012) Moving from static to dynamic complexity in hydrogel design. Nat Commun 3:1269
48. Guvendiren M, Burdick JA (2012) Stiffening hydrogels to probe short- and long-term cellular responses to dynamic mechanics. Nat Commun 3:792. https://doi.org/10.1038/ncomms1792
49. Khademhosseini A, Langer R, Borenstein J, Vacanti JP (2006) Microscale technologies for tissue engineering and biology. Proc Natl Acad Sci 103:2480–2487. https://doi.org/10.1073/pnas.0507681102
50. Annabi N, Tamayol A, Uquillas JA et al (2014) 25th anniversary article: rational design and applications of hydrogels in regenerative medicine. Adv Mater 26:85–124. https://doi.org/10.1002/adma.201303233
51. Albrecht DR, Underhill GH, Wassermann TB et al (2006) Probing the role of multicellular organization in three-dimensional microenvironments. Nat Methods 3:369–375. https://doi.org/10.1038/nmeth873
52. Du Y, Lo E, Ali S, Khademhosseini A (2008) Directed assembly of cell-laden microgels for fabrication of 3D tissue constructs. Proc Natl Acad Sci 105:9522–9527. https://doi.org/10.1073/pnas.0801866105
53. Gartner ZJ, Bertozzi CR (2009) Programmed assembly of 3-dimensional microtissues with defined cellular connectivity. Proc Natl Acad Sci 106:4606–4610. https://doi.org/10.1073/pnas.0900717106
54. Nichol JW, Khademhosseini A (2009) Modular tissue engineering: engineering biological tissues from the bottom up. Soft Matter 5:1312–1319. https://doi.org/10.1039/B814285H
55. Sun W, Starly B, Darling A, Gomez C (2004) Computer-aided tissue engineering: application to biomimetic modelling and design of tissue scaffolds. Biotechnol Appl Biochem 39:49–58. https://doi.org/10.1042/BA20030109
56. Ang TH, Sultana FSA, Hutmacher DW et al (2002) Fabrication of 3D chitosan–hydroxyapatite scaffolds using a robotic dispensing system. Mater Sci Eng C 20:35–42. https://doi.org/10.1016/S0928-4931(02)00010-3
57. Landers R, Hübner U, Schmelzeisen R, Mülhaupt R (2002) Rapid prototyping of scaffolds derived from thermoreversible hydrogels and tailored for applications in tissue engineering. Biomaterials 23:4437–4447
58. Boland T, Mironov V, Gutowska A et al (2003) Cell and organ printing 2: fusion of cell aggregates in three-dimensional gels. Anat Rec A Discov Mol Cell Evol Biol 272A:497–502. https://doi.org/10.1002/ar.a.10059
59. Mironov V, Visconti RP, Kasyanov V et al (2009) Organ printing: tissue spheroids as building blocks. Biomaterials 30:2164–2174
60. Marquez GJ, Renn MJ, Miller WD (2001) Aerosol-based direct-write of biological materials for biomedical applications. Mat Res Soc Symp Proc Arch 698:343–349. https://doi.org/10.1557/PROC-698-Q5.2.1

61. Ringeisen BR, Kim H, Young HD et al (2001) Cell-by-cell construction of living tissue. Mat Res Soc Symp Proc Arch 698. https://doi.org/10.1557/PROC-698-Q5.1.1
62. Ringeisen BR, Chrisey DB, Krizman DB et al (2002) Cell-by-cell construction of living tissue by ambient laser transfer. In: Second annual international IEEE-EMBS special topic conference on microtechnologies in medicine and biology. Proceedings (Cat. No.02EX578), pp 120–125
63. Kachurin AM, Stewart RL, Church KH et al (2001) Direct-write construction of tissue-engineered scaffolds. Mat Res Soc Symp Proc Arch 698:1–6. https://doi.org/10.1557/PROC-698-Q5.5.1
64. Smith CM, Stone AL, Parkhill RL et al (2004) Three-dimensional bioassembly tool for generating viable tissue-engineered constructs. Tissue Eng 10:1566–1576. https://doi.org/10.1089/ten.2004.10.1566
65. Hutmacher DW, Sittinger M, Risbud MV (2004) Scaffold-based tissue engineering: rationale for computer-aided design and solid free-form fabrication systems. Trends Biotechnol 22:354–362. https://doi.org/10.1016/j.tibtech.2004.05.005
66. Khalil S, Nam J, Sun W (2005) Multi-nozzle deposition for construction of 3D biopolymer tissue scaffolds. Rapid Prototyp J 11:9–17. https://doi.org/10.1108/13552540510573347
67. Cohen DL, Malone E, Lipson H, Bonassar LJ (2006) Direct freeform fabrication of seeded hydrogels in arbitrary geometries. Tissue Eng 12:1325–1335. https://doi.org/10.1089/ten.2006.12.1325
68. Liu VA, Bhatia SN (2002) Three-dimensional photopatterning of hydrogels containing living cells. Biomed Microdevices 4:257–266. https://doi.org/10.1023/A:1020932105236
69. Yu T, Chiellini F, Schmaljohan D et al (2002) Microfabrication of hydrogels for biomedical applications. In: Advances in resist technology and processing XIX. International Society for Optics and Photonics, pp 854–861
70. Dhariwala B, Hunt E, Boland T (2004) Rapid prototyping of tissue-engineering constructs, using photopolymerizable hydrogels and stereolithography. Tissue Eng 10:1316–1322. https://doi.org/10.1089/ten.2004.10.1316
71. Tayalia P, Mendonca CR, Baldacchini T et al (2008) 3D cell-migration studies using two-photon engineered polymer scaffolds. Adv Mater 20:4494–4498. https://doi.org/10.1002/adma.200801319
72. Hahn MS, Miller JS, West JL (2006) Three-dimensional biochemical and biomechanical patterning of hydrogels for guiding cell behavior. Adv Mater 18:2679–2684. https://doi.org/10.1002/adma.200600647
73. Lee S-H, Moon JJ, West JL (2008) Three-dimensional micropatterning of bioactive hydrogels via two-photon laser scanning photolithography for guided 3D cell migration. Biomaterials 29:2962–2968. https://doi.org/10.1016/j.biomaterials.2008.04.004
74. Xing J-F, Zheng M-L, Duan X-M (2015) Two-photon polymerization microfabrication of hydrogels: an advanced 3D printing technology for tissue engineering and drug delivery. Chem Soc Rev 44:5031–5039. https://doi.org/10.1039/C5CS00278H
75. Guvendiren M, Burdick JA (2013) Engineering synthetic hydrogel microenvironments to instruct stem cells. Curr Opin Biotechnol 24:841–846. https://doi.org/10.1016/j.copbio.2013.03.009
76. Mironov V, Reis N, Derby B (2006) Review: bioprinting: a beginning. Tissue Eng 12:631–634. https://doi.org/10.1089/ten.2006.12.631
77. Derby B (2012) Printing and prototyping of tissues and scaffolds. Science 338:921–926. https://doi.org/10.1126/science.1226340
78. Xu T, Kincaid H, Atala A, Yoo JJ (2008) High-throughput production of single-cell microparticles using an inkjet printing technology. J Manuf Sci Eng 130:021017. https://doi.org/10.1115/1.2903064
79. Xu T, Olson J, Zhao W et al (2008) Characterization of cell constructs generated with inkjet printing technology using in vivo magnetic resonance imaging. J Manuf Sci Eng 130:021013. https://doi.org/10.1115/1.2902857

80. Malda J, Visser J, Melchels FP et al (2013) 25th anniversary article: engineering hydrogels for biofabrication. Adv Mater 25:5011–5028. https://doi.org/10.1002/adma.201302042
81. Nakamura M, Iwanaga S, Henmi C et al (2010) Biomatrices and biomaterials for future developments of bioprinting and biofabrication. Biofabrication 2:014110. https://doi.org/10.1088/1758-5082/2/1/014110
82. Wang X, Yan Y, Pan Y et al (2006) Generation of three-dimensional hepatocyte/gelatin structures with rapid prototyping system. Tissue Eng 12:83–90. https://doi.org/10.1089/ten.2006.12.83
83. Jones N (2012) Science in three dimensions: the print revolution. Nature 487:22. https://doi.org/10.1038/487022a
84. Brown TD, Dalton PD, Hutmacher DW (2011) Direct writing by way of melt electrospinning. Adv Mater 23:5651–5657. https://doi.org/10.1002/adma.201103482
85. Dalton PD (2017) Melt electrowriting with additive manufacturing principles. Curr Opin Biomed Eng 2:49–57. https://doi.org/10.1016/j.cobme.2017.05.007
86. Khademhosseini A, Langer R (2016) A decade of progress in tissue engineering. Nat Protoc 11:1775–1781. https://doi.org/10.1038/nprot.2016.123
87. Groll J, Burdick JA, Cho D-W et al (2018) A definition of bioinks and their distinction from biomaterial inks. Biofabrication 11:013001. https://doi.org/10.1088/1758-5090/aaec52
88. Bandyopadhyay A, Bose S, Das S (2015) 3D printing of biomaterials [refer to the sidebar by Hutmacher in print version of this article]. MRS Bull 40:108–115. https://doi.org/10.1557/mrs.2015.3
89. Guillemot F, Mironov V, Nakamura M (2010) Bioprinting is coming of age: report from the International Conference on Bioprinting and Biofabrication in Bordeaux (3B\textquotesingle09). Biofabrication 2:010201. https://doi.org/10.1088/1758-5082/2/1/010201
90. Elbert DL (2011) Bottom-up tissue engineering. Curr Opin Biotechnol 22:674–680. https://doi.org/10.1016/j.copbio.2011.04.001
91. Guven S, Chen P, Inci F et al (2015) Multiscale assembly for tissue engineering and regenerative medicine. Trends Biotechnol 33:269–279. https://doi.org/10.1016/j.tibtech.2015.02.003
92. Guvendiren M, Molde J, Soares RM, Kohn J (2016) Designing biomaterials for 3D printing. ACS Biomater Sci Eng 2:1679–1693
93. Cox SC, Thornby JA, Gibbons GJ et al (2015) 3D printing of porous hydroxyapatite scaffolds intended for use in bone tissue engineering applications. Mater Sci Eng C 47:237–247. https://doi.org/10.1016/j.msec.2014.11.024
94. Hospodiuk M, Dey M, Sosnoski D, Ozbolat IT (2017) The bioink: a comprehensive review on bioprintable materials. Biotechnol Adv 35:217–239. https://doi.org/10.1016/j.biotechadv.2016.12.006
95. Ferris CJ, Gilmore KJ, Beirne S et al (2013) Bio-ink for on-demand printing of living cells. Biomater Sci 1:224–230. https://doi.org/10.1039/C2BM00114D
96. Mironov V (2003) Printing technology to produce living tissue. Expert Opin Biol Ther 3:701–704. https://doi.org/10.1517/14712598.3.5.701
97. Mironov V, Boland T, Trusk T et al (2003) Organ printing: computer-aided jet-based 3D tissue engineering. Trends Biotechnol 21:157–161. https://doi.org/10.1016/S0167-7799(03)00033-7
98. Keriquel V, Guillemot F, Arnault I et al (2010) In vivo bioprinting for computer- and robotic-assisted medical intervention: preliminary study in mice. Biofabrication 2:014101. https://doi.org/10.1088/1758-5082/2/1/014101
99. Moroni L, Boland T, Burdick JA et al (2018) Biofabrication: a guide to technology and terminology. Trends Biotechnol 36:384–402. https://doi.org/10.1016/j.tibtech.2017.10.015
100. Bücking TM, Hill ER, Robertson JL et al (2017) From medical imaging data to 3D printed anatomical models. PLoS One 12:e0178540. https://doi.org/10.1371/journal.pone.0178540
101. Leukers B, Gülkan H, Irsen SH et al (2005) Hydroxyapatite scaffolds for bone tissue engineering made by 3D printing. J Mater Sci Mater Med 16:1121–1124. https://doi.org/10.1007/s10856-005-4716-5

102. Murphy SV, Atala A (2014) 3D bioprinting of tissues and organs. Nat Biotechnol 32:773–785. https://doi.org/10.1038/nbt.2958
103. Wang LV, Hu S (2012) Photoacoustic tomography: in vivo imaging from organelles to organs. Science 335:1458–1462. https://doi.org/10.1126/science.1216210
104. Maslov K, Zhang HF, Hu S, Wang LV (2008) Optical-resolution photoacoustic microscopy for in vivo imaging of single capillaries. Opt Lett 33:929–931. https://doi.org/10.1364/OL.33.000929
105. Peltola SM, Melchels FPW, Grijpma DW, Kellomäki M (2008) A review of rapid prototyping techniques for tissue engineering purposes. Ann Med 40:268–280. https://doi.org/10.1080/07853890701881788
106. Kinstlinger IS, Bastian A, Paulsen SJ et al (2016) Open-source selective laser sintering (OpenSLS) of nylon and biocompatible polycaprolactone. PLoS One 11:e0147399. https://doi.org/10.1371/journal.pone.0147399
107. Skardal A, Zhang J, McCoard L et al (2010) Photocrosslinkable hyaluronan-gelatin hydrogels for two-step bioprinting. Tissue Eng A 16:2675–2685. https://doi.org/10.1089/ten.tea.2009.0798
108. Boere KWM, Blokzijl MM, Visser J et al (2015) Biofabrication of reinforced 3D-scaffolds using two-component hydrogels. J Mater Chem B 3:9067–9078. https://doi.org/10.1039/C5TB01645B
109. Rutz AL, Hyland KE, Jakus AE et al (2015) A multimaterial bioink method for 3D printing tunable, cell-compatible hydrogels. Adv Mater 27:1607–1614
110. Ouyang L, Highley CB, Rodell CB et al (2016) 3D printing of shear-thinning hyaluronic acid hydrogels with secondary cross-linking. ACS Biomater Sci Eng 2:1743–1751
111. Highley CB, Song KH, Daly AC, Burdick JA (2019) Jammed microgel inks for 3D printing applications. Adv Sci 6:1801076. https://doi.org/10.1002/advs.201801076
112. Bhattacharjee T, Zehnder SM, Rowe KG et al (2015) Writing in the granular gel medium. Sci Adv 1:e1500655
113. Hinton TJ, Jallerat Q, Palchesko RN et al (2015) Three-dimensional printing of complex biological structures by freeform reversible embedding of suspended hydrogels. Sci Adv 1:e1500758
114. Hockaday LA, Kang KH, Colangelo NW et al (2012) Rapid 3D printing of anatomically accurate and mechanically heterogeneous aortic valve hydrogel scaffolds. Biofabrication 4:035005. https://doi.org/10.1088/1758-5082/4/3/035005
115. Phillippi JA, Miller E, Weiss L et al (2008) Microenvironments engineered by inkjet bioprinting spatially direct adult stem cells toward muscle- and bone-like subpopulations. Stem Cells 26:127–134. https://doi.org/10.1634/stemcells.2007-0520
116. Laronda MM, Rutz AL, Xiao S et al (2017) A bioprosthetic ovary created using 3D printed microporous scaffolds restores ovarian function in sterilized mice. Nat Commun 8:15261. https://doi.org/10.1038/ncomms15261
117. Guvendiren M, Fung S, Kohn J et al (2017) The control of stem cell morphology and differentiation using three-dimensional printed scaffold architecture. MRS Commun 7:383–390. https://doi.org/10.1557/mrc.2017.73
118. Song KH, Highley CB, Rouff A, Burdick JA (2018) Complex 3D-printed microchannels within cell-degradable hydrogels. Adv Funct Mater 28:1801331. https://doi.org/10.1002/adfm.201801331
119. Datta P, Ayan B, Ozbolat IT (2017) Bioprinting for vascular and vascularized tissue biofabrication. Acta Biomater 51:1–20. https://doi.org/10.1016/j.actbio.2017.01.035
120. Koike N, Fukumura D, Gralla O et al (2004) Tissue engineering: creation of long-lasting blood vessels. Nature 428:138–139. https://doi.org/10.1038/428138a
121. Cuchiara MP, Gould DJ, MK MH et al (2012) Integration of self-assembled microvascular networks with microfabricated PEG-based hydrogels. Adv Funct Mater 22:4511–4518. https://doi.org/10.1002/adfm.201200976

122. Campisi M, Shin Y, Osaki T et al (2018) 3D self-organized microvascular model of the human blood-brain barrier with endothelial cells, pericytes and astrocytes. Biomaterials 180:117–129. https://doi.org/10.1016/j.biomaterials.2018.07.014
123. Thilmany JA. Robot that prints tissue. https://www.asme.org/engineering-topics/articles/bioengineering/a-robot-that-prints-tissue. Accessed 16 Mar 2019
124. Miller JS, Stevens KR, Yang MT et al (2012) Rapid casting of patterned vascular networks for perfusable engineered 3D tissues. Nat Mater 11:768
125. Kolesky DB, Homan KA, Skylar-Scott MA, Lewis JA (2016) Three-dimensional bioprinting of thick vascularized tissues. Proc Natl Acad Sci 113:3179–3184
126. Highley CB, Rodell CB, Burdick JA (2015) Direct 3D printing of shear-thinning hydrogels into self-healing hydrogels. Adv Mater 27:5075–5079
127. Daly AC, Pitacco P, Nulty J et al (2018) 3D printed microchannel networks to direct vascularisation during endochondral bone repair. Biomaterials 162:34–46. https://doi.org/10.1016/j.biomaterials.2018.01.057
128. Ozbolat IT, Moncal KK, Gudapati H (2017) Evaluation of bioprinter technologies. Addit Manuf 13:179–200. https://doi.org/10.1016/j.addma.2016.10.003
129. Ozbolat IT, Hospodiuk M (2016) Current advances and future perspectives in extrusion-based bioprinting. Biomaterials 76:321–343. https://doi.org/10.1016/j.biomaterials.2015.10.076
130. Khalil S, Sun W (2007) Biopolymer deposition for freeform fabrication of hydrogel tissue constructs. Mater Sci Eng C 27:469–478. https://doi.org/10.1016/j.msec.2006.05.023
131. Chang CC, Boland ED, Williams SK, Hoying JB (2011) Direct-write bioprinting three-dimensional biohybrid systems for future regenerative therapies. J Biomed Mater Res B Appl Biomater 98B:160–170. https://doi.org/10.1002/jbm.b.31831
132. Jakab K, Damon B, Neagu A et al (2006) Three-dimensional tissue constructs built by bioprinting. Biorheology 43:509–513
133. Visser J, Peters B, Burger TJ et al (2013) Biofabrication of multi-material anatomically shaped tissue constructs. Biofabrication 5:035007. https://doi.org/10.1088/1758-5082/5/3/035007
134. Xiong Z, Yan Y, Wang S et al (2002) Fabrication of porous scaffolds for bone tissue engineering via low-temperature deposition. Scr Mater 46:771–776. https://doi.org/10.1016/S1359-6462(02)00071-4
135. Vozzi G, Ahluwalia A (2007) Microfabrication for tissue engineering: rethinking the cells-on-a scaffold approach. J Mater Chem 17:1248–1254. https://doi.org/10.1039/B613511K
136. Jakab K, Norotte C, Marga F et al (2010) Tissue engineering by self-assembly and bio-printing of living cells. Biofabrication 2:022001. https://doi.org/10.1088/1758-5082/2/2/022001
137. Yeong W-Y, Chua C-K, Leong K-F, Chandrasekaran M (2004) Rapid prototyping in tissue engineering: challenges and potential. Trends Biotechnol 22:643–652. https://doi.org/10.1016/j.tibtech.2004.10.004
138. Tan Z, Parisi C, Silvio LD et al (2017) Cryogenic 3D printing of super soft hydrogels. Sci Rep 7:16293. https://doi.org/10.1038/s41598-017-16668-9
139. Vozzi G, Previti A, De Rossi D, Ahluwalia A (2002) Microsyringe-based deposition of two-dimensional and three-dimensional polymer scaffolds with a well-defined geometry for application to tissue engineering. Tissue Eng 8:1089–1098. https://doi.org/10.1089/107632702320934182
140. Gratson GM, Xu M, Lewis JA (2004) Microperiodic structures: direct writing of three-dimensional webs. Nature 428:386. https://doi.org/10.1038/428386a
141. Gladman AS, Matsumoto EA, Nuzzo RG et al (2016) Biomimetic 4D printing. Nat Mater 15:413–418. https://doi.org/10.1038/nmat4544
142. Jakab K, Neagu A, Mironov V et al (2004) Engineering biological structures of prescribed shape using self-assembling multicellular systems. Proc Natl Acad Sci 101:2864–2869. https://doi.org/10.1073/pnas.0400164101
143. Norotte C, Marga FS, Niklason LE, Forgacs G (2009) Scaffold-free vascular tissue engineering using bioprinting. Biomaterials 30:5910–5917. https://doi.org/10.1016/j.biomaterials.2009.06.034

144. Marga F, Jakab K, Khatiwala C et al (2012) Toward engineering functional organ modules by additive manufacturing. Biofabrication 4:022001. https://doi.org/10.1088/1758-5082/4/2/022001
145. Yu Y, Moncal KK, Li J et al (2016) Three-dimensional bioprinting using self-assembling scalable scaffold-free "tissue strands" as a new bioink. Sci Rep 6:28714. https://doi.org/10.1038/srep28714
146. Pfister A, Landers R, Laib A et al (2004) Biofunctional rapid prototyping for tissue-engineering applications: 3D bioplotting versus 3D printing. J Polym Sci Part Polym Chem 42:624–638. https://doi.org/10.1002/pola.10807
147. Jakab K, Norotte C, Damon B et al (2008) Tissue engineering by self-assembly of cells printed into topologically defined structures. Tissue Eng A 14:413–421. https://doi.org/10.1089/tea.2007.0173
148. Moldovan NI, Hibino N, Nakayama K (2016) Principles of the Kenzan method for robotic cell spheroid-based three-dimensional bioprinting. Tissue Eng B Rev 23:237–244. https://doi.org/10.1089/ten.teb.2016.0322
149. Mao T, Kuhn DCS, Tran H (1997) Spread and rebound of liquid droplets upon impact on flat surfaces. AICHE J 43:2169–2179. https://doi.org/10.1002/aic.690430903
150. Derby B (2010) Inkjet printing of functional and structural materials: fluid property requirements, feature stability, and resolution. Annu Rev Mater Res 40:395–414. https://doi.org/10.1146/annurev-matsci-070909-104502
151. Jang D, Kim D, Moon J (2009) Influence of fluid physical properties on ink-jet printability. Langmuir 25:2629–2635. https://doi.org/10.1021/la900059m
152. Gudapati H, Dey M, Ozbolat I (2016) A comprehensive review on droplet-based bioprinting: past, present and future. Biomaterials 102:20–42. https://doi.org/10.1016/j.biomaterials.2016.06.012
153. Derby B (2008) Bioprinting: inkjet printing proteins and hybrid cell-containing materials and structures. J Mater Chem 18:5717–5721. https://doi.org/10.1039/B807560C
154. Stringer J, Derby B (2009) Limits to feature size and resolution in ink jet printing. J Eur Ceram Soc 29:913–918. https://doi.org/10.1016/j.jeurceramsoc.2008.07.016
155. van Osch THJ, Perelaer J, de Laat AWM, Schubert US (2008) Inkjet printing of narrow conductive tracks on untreated polymeric substrates. Adv Mater 20:343–345. https://doi.org/10.1002/adma.200701876
156. Xu T, Jin J, Gregory C et al (2005) Inkjet printing of viable mammalian cells. Biomaterials 26:93–99. https://doi.org/10.1016/j.biomaterials.2004.04.011
157. Xu T, Gregory CA, Molnar P et al (2006) Viability and electrophysiology of neural cell structures generated by the inkjet printing method. Biomaterials 27:3580–3588. https://doi.org/10.1016/j.biomaterials.2006.01.048
158. Cui X, Dean D, Ruggeri ZM, Boland T (2010) Cell damage evaluation of thermal inkjet printed Chinese hamster ovary cells. Biotechnol Bioeng 106:963–969. https://doi.org/10.1002/bit.22762
159. Saunders RE, Gough JE, Derby B (2008) Delivery of human fibroblast cells by piezoelectric drop-on-demand inkjet printing. Biomaterials 29:193–203. https://doi.org/10.1016/j.biomaterials.2007.09.032
160. Xu C, Chai W, Huang Y, Markwald RR (2012) Scaffold-free inkjet printing of three-dimensional zigzag cellular tubes. Biotechnol Bioeng 109:3152–3160. https://doi.org/10.1002/bit.24591
161. Christensen K, Xu C, Chai W et al (2015) Freeform inkjet printing of cellular structures with bifurcations. Biotechnol Bioeng 112:1047–1055. https://doi.org/10.1002/bit.25501
162. Xu T, Baicu C, Aho M et al (2009) Fabrication and characterization of bio-engineered cardiac pseudo tissues. Biofabrication 1:035001. https://doi.org/10.1088/1758-5082/1/3/035001
163. Campbell PG, Weiss LE (2007) Tissue engineering with the aid of inkjet printers. Expert Opin Biol Ther 7:1123–1127. https://doi.org/10.1517/14712598.7.8.1123
164. Long Ng W, Min Lee J, Yee Yeong W, Naing MW (2017) Microvalve-based bioprinting—process, bio-inks and applications. Biomater Sci 5:632–647. https://doi.org/10.1039/C6BM00861E

165. Moon S, Hasan SK, Song YS et al (2009) Layer by layer three-dimensional tissue epitaxy by cell-laden hydrogel droplets. Tissue Eng C Methods 16:157–166. https://doi.org/10.1089/ten.tec.2009.0179
166. Gruene M, Unger C, Koch L et al (2011) Dispensing pico to nanolitre of a natural hydrogel by laser-assisted bioprinting. Biomed Eng Online 10:19. https://doi.org/10.1186/1475-925X-10-19
167. Elrod SA, Hadimioglu B, Khuri-Yakub BT et al (1989) Nozzleless droplet formation with focused acoustic beams. J Appl Phys 65:3441–3447. https://doi.org/10.1063/1.342663
168. Fang Y, Frampton JP, Raghavan S et al (2012) Rapid generation of multiplexed cell cocultures using acoustic droplet ejection followed by aqueous two-phase exclusion patterning. Tissue Eng C Methods 18:647–657. https://doi.org/10.1089/ten.tec.2011.0709
169. Onses MS, Sutanto E, Ferreira PM et al (2015) Mechanisms, capabilities, and applications of high-resolution electrohydrodynamic jet printing. Small 11:4237–4266. https://doi.org/10.1002/smll.201500593
170. Lee W, Debasitis JC, Lee VK et al (2009) Multi-layered culture of human skin fibroblasts and keratinocytes through three-dimensional freeform fabrication. Biomaterials 30:1587–1595. https://doi.org/10.1016/j.biomaterials.2008.12.009
171. Chahal D, Ahmadi A, Cheung KC (2012) Improving piezoelectric cell printing accuracy and reliability through neutral buoyancy of suspensions. Biotechnol Bioeng 109:2932–2940. https://doi.org/10.1002/bit.24562
172. Demirci U, Montesano G (2007) Cell encapsulating droplet vitrification. Lab Chip 7:1428–1433. https://doi.org/10.1039/B705809H
173. Demirci U, Montesano G (2007) Single cell epitaxy by acoustic picolitre droplets. Lab Chip 7:1139–1145. https://doi.org/10.1039/B704965J
174. Demirci U (2006) Acoustic picoliter droplets for emerging applications in semiconductor industry and biotechnology. J Microelectromech Syst 15:957–966. https://doi.org/10.1109/JMEMS.2006.878879
175. Jayasinghe SN, Qureshi AN, Eagles PAM (2006) Electrohydrodynamic jet processing: an advanced electric-field-driven jetting phenomenon for processing living cells. Small 2:216–219. https://doi.org/10.1002/smll.200500291
176. Wade RJ, Burdick JA (2014) Advances in nanofibrous scaffolds for biomedical applications: from electrospinning to self-assembly. Nano Today 9:722–742. https://doi.org/10.1016/j.nantod.2014.10.002
177. Gasperini L, Maniglio D, Motta A, Migliaresi C (2014) An electrohydrodynamic bioprinter for alginate hydrogels containing living cells. Tissue Eng C Methods 21:123–132. https://doi.org/10.1089/ten.tec.2014.0149
178. Hayati I, Bailey AI, Tadros TF (1986) Mechanism of stable jet formation in electrohydrodynamic atomization. Nature 319:41. https://doi.org/10.1038/319041a0
179. Eagles PAM, Qureshi AN, Jayasinghe SN (2006) Electrohydrodynamic jetting of mouse neuronal cells. Biochem J 394:375–378. https://doi.org/10.1042/BJ20051838
180. Workman VL, Tezera LB, Elkington PT, Jayasinghe SN (2014) Controlled generation of microspheres incorporating extracellular matrix fibrils for three-dimensional cell culture. Adv Funct Mater 24:2648–2657. https://doi.org/10.1002/adfm.201303891
181. Xie J, Wang C-H (2007) Electrospray in the dripping mode for cell microencapsulation. J Colloid Interface Sci 312:247–255. https://doi.org/10.1016/j.jcis.2007.04.023
182. Poellmann MJ, Barton KL, Mishra S, Johnson AJW (2011) Patterned hydrogel substrates for cell culture with electrohydrodynamic jet printing. Macromol Biosci 11:1164–1168. https://doi.org/10.1002/mabi.201100004
183. Barron JA, Wu P, Ladouceur HD, Ringeisen BR (2004) Biological laser printing: a novel technique for creating heterogeneous 3-dimensional cell patterns. Biomed Microdevices 6:139–147. https://doi.org/10.1023/B:BMMD.0000031751.67267.9f
184. Koch L, Kuhn S, Sorg H et al (2009) Laser printing of skin cells and human stem cells. Tissue Eng C Methods 16:847–854. https://doi.org/10.1089/ten.tec.2009.0397

185. Guillotin B, Souquet A, Catros S et al (2010) Laser assisted bioprinting of engineered tissue with high cell density and microscale organization. Biomaterials 31:7250–7256. https://doi.org/10.1016/j.biomaterials.2010.05.055
186. Barron JA, Ringeisen BR, Kim H et al (2004) Application of laser printing to mammalian cells. Thin Solid Films 453–454:383–387. https://doi.org/10.1016/j.tsf.2003.11.161
187. Mézel C, Hallo L, Souquet A et al (2009) Self-consistent modeling of jet formation process in the nanosecond laser pulse regime. Phys Plasmas 16:123112. https://doi.org/10.1063/1.3276101
188. Mézel C, Souquet A, Hallo L, Guillemot F (2010) Bioprinting by laser-induced forward transfer for tissue engineering applications: jet formation modeling. Biofabrication 2:014103. https://doi.org/10.1088/1758-5082/2/1/014103
189. Yap CY, Chua CK, Dong ZL et al (2015) Review of selective laser melting: materials and applications. Appl Phys Rev 2:041101. https://doi.org/10.1063/1.4935926
190. Rombouts M, Kruth J, Froyen L et al (2005) Binding mechanisms in selective laser sintering and selective laser melting. Rapid Prototyp J 11:26–36. https://doi.org/10.1108/13552540510573365
191. Melchels FPW, Feijen J, Grijpma DW (2010) A review on stereolithography and its applications in biomedical engineering. Biomaterials 31:6121–6130. https://doi.org/10.1016/j.biomaterials.2010.04.050
192. Bückmann T, Stenger N, Kadic M et al (2012) Tailored 3D mechanical metamaterials made by dip-in direct-laser-writing optical lithography. Adv Mater 24:2710–2714. https://doi.org/10.1002/adma.201200584
193. Lee S-J, Kang H-W, Park JK et al (2008) Application of microstereolithography in the development of three-dimensional cartilage regeneration scaffolds. Biomed Microdevices 10:233–241. https://doi.org/10.1007/s10544-007-9129-4
194. Hribar KC, Soman P, Warner J et al (2014) Light-assisted direct-write of 3D functional biomaterials. Lab Chip 14:268–275. https://doi.org/10.1039/C3LC50634G
195. Zhang AP, Qu X, Soman P et al (2012) Rapid fabrication of complex 3D extracellular microenvironments by dynamic optical projection stereolithography. Adv Mater 24:4266–4270
196. Sun C, Fang N, Wu DM, Zhang X (2005) Projection micro-stereolithography using digital micro-mirror dynamic mask. Sens Actuators Phys 121:113–120. https://doi.org/10.1016/j.sna.2004.12.011
197. Lu Y, Mapili G, Suhali G et al (2006) A digital micro-mirror device-based system for the microfabrication of complex, spatially patterned tissue engineering scaffolds. J Biomed Mater Res A 77A:396–405. https://doi.org/10.1002/jbm.a.30601
198. Lim KS, Levato R, Costa PF et al (2018) Bio-resin for high resolution lithography-based biofabrication of complex cell-laden constructs. Biofabrication 10:034101. https://doi.org/10.1088/1758-5090/aac00c
199. Tumbleston JR, Shirvanyants D, Ermoshkin N et al (2015) Continuous liquid interface production of 3D objects. Science 347:1349–1352. https://doi.org/10.1126/science.aaa2397
200. Lin H, Zhang D, Alexander PG et al (2013) Application of visible light-based projection stereolithography for live cell-scaffold fabrication with designed architecture. Biomaterials 34:331–339. https://doi.org/10.1016/j.biomaterials.2012.09.048
201. Kim S, Qiu F, Kim S et al (2013) Fabrication and characterization of magnetic microrobots for three-dimensional cell culture and targeted transportation. Adv Mater 25:5863–5868. https://doi.org/10.1002/adma.201301484
202. Greiner AM, Klein F, Gudzenko T et al (2015) Cell type-specific adaptation of cellular and nuclear volume in micro-engineered 3D environments. Biomaterials 69:121–132. https://doi.org/10.1016/j.biomaterials.2015.08.016
203. Kloxin AM, Kasko AM, Salinas CN, Anseth KS (2009) Photodegradable hydrogels for dynamic tuning of physical and chemical properties. Science 324:59–63. https://doi.org/10.1126/science.1169494
204. Arakawa CK, Badeau BA, Zheng Y, DeForest CA (2017) Multicellular vascularized engineered tissues through user-programmable biomaterial photodegradation. Adv Mater 29(37):1703156

205. Brandenberg N, Lutolf MP (2016) In situ patterning of microfluidic networks in 3D cell-laden hydrogels. Adv Mater 28:7450–7456
206. DeForest CA, Polizzotti BD, Anseth KS (2009) Sequential click reactions for synthesizing and patterning three-dimensional cell microenvironments. Nat Mater 8:659–664. https://doi.org/10.1038/nmat2473
207. Zhou X, Hou Y, Lin J (2015) A review on the processing accuracy of two-photon polymerization. AIP Adv 5:030701. https://doi.org/10.1063/1.4916886
208. Xing J-F, Dong X-Z, Chen W-Q et al (2007) Improving spatial resolution of two-photon microfabrication by using photoinitiator with high initiating efficiency. Appl Phys Lett 90:131106. https://doi.org/10.1063/1.2717532
209. Fischer J, von Freymann G, Wegener M (2010) The materials challenge in diffraction-unlimited direct-laser-writing optical lithography. Adv Mater 22:3578–3582. https://doi.org/10.1002/adma.201000892
210. Della Giustina G, Gandin A, Brigo L et al (2019) Polysaccharide hydrogels for multiscale 3D printing of pullulan scaffolds. Mater Des 165:107566. https://doi.org/10.1016/j.matdes.2018.107566
211. Nguyen AK, Narayan RJ (2017) Two-photon polymerization for biological applications. Mater Today 20:314–322. https://doi.org/10.1016/j.mattod.2017.06.004
212. Accardo A, Blatché M-C, Courson R et al (2018) Two-photon lithography and microscopy of 3D hydrogel scaffolds for neuronal cell growth. Biomed Phys Eng Exp 4:027009. https://doi.org/10.1088/2057-1976/aaab93
213. Brigo L, Urciuolo A, Giulitti S et al (2017) 3D high-resolution two-photon crosslinked hydrogel structures for biological studies. Acta Biomater 55:373–384. https://doi.org/10.1016/j.actbio.2017.03.036
214. Schuurman W, Khristov V, Pot MW et al (2011) Bioprinting of hybrid tissue constructs with tailorable mechanical properties. Biofabrication 3:021001. https://doi.org/10.1088/1758-5082/3/2/021001
215. Mekhileri NV, Lim KS, Brown GCJ et al (2018) Automated 3D bioassembly of micro-tissues for biofabrication of hybrid tissue engineered constructs. Biofabrication 10:024103. https://doi.org/10.1088/1758-5090/aa9ef1
216. Jungst T, Smolan W, Schacht K et al (2016) Strategies and molecular design criteria for 3D printable hydrogels. Chem Rev 116:1496–1539. https://doi.org/10.1021/acs.chemrev.5b00303
217. Dubbin K, Hori Y, Lewis KK, Heilshorn SC (2016) Dual-stage crosslinking of a gel-phase bioink improves cell viability and homogeneity for 3D bioprinting. Adv Healthc Mater 5:2488–2492
218. Ouyang L, Highley CB, Sun W, Burdick JA (2017) A generalizable strategy for the 3D bioprinting of hydrogels from nonviscous photo-crosslinkable inks. Adv Mater 29, 1604983
219. Chang R, Nam J, Sun W (2008) Effects of dispensing pressure and nozzle diameter on cell survival from solid freeform fabrication–based direct cell writing. Tissue Eng A 14:41–48. https://doi.org/10.1089/ten.a.2007.0004
220. Aguado BA, Mulyasasmita W, Su J et al (2011) Improving viability of stem cells during syringe needle flow through the design of hydrogel cell carriers. Tissue Eng A 18:806–815. https://doi.org/10.1089/ten.tea.2011.0391
221. Nair K, Gandhi M, Khalil S et al (2009) Characterization of cell viability during bioprinting processes. Biotechnol J 4:1168–1177. https://doi.org/10.1002/biot.200900004
222. Hölzl K, Lin S, Tytgat L et al (2016) Bioink properties before, during and after 3D bioprinting. Biofabrication 8:032002. https://doi.org/10.1088/1758-5090/8/3/032002
223. Colosi C, Shin SR, Manoharan V et al (2016) Microfluidic bioprinting of heterogeneous 3D tissue constructs using low-viscosity bioink. Adv Mater 28:677–684
224. Miller JS (2014) The billion cell construct: will three-dimensional printing get us there? PLoS Biol 12:e1001882. https://doi.org/10.1371/journal.pbio.1001882
225. Mendicino M, Bailey AM, Wonnacott K et al (2014) MSC-based product characterization for clinical trials: an FDA perspective. Cell Stem Cell 14:141–145. https://doi.org/10.1016/j.stem.2014.01.013

226. Marklein RA, Lam J, Guvendiren M et al (2018) Functionally-relevant morphological profiling: a tool to assess cellular heterogeneity. Trends Biotechnol 36:105–118. https://doi.org/10.1016/j.tibtech.2017.10.007
227. Simaria AS, Hassan S, Varadaraju H et al (2014) Allogeneic cell therapy bioprocess economics and optimization: single-use cell expansion technologies. Biotechnol Bioeng 111:69–83. https://doi.org/10.1002/bit.25008
228. Vacanti JP, Morse MA, Saltzman WM et al (1988) Selective cell transplantation using bioabsorbable artificial polymers as matrices. J Pediatr Surg 23:3–9. https://doi.org/10.1016/S0022-3468(88)80529-3
229. Saunders RE, Derby B (2014) Inkjet printing biomaterials for tissue engineering: bioprinting. Int Mater Rev 59:430–448. https://doi.org/10.1179/1743280414Y.0000000040
230. Nakamura M, Kobayashi A, Takagi F et al (2005) Biocompatible inkjet printing technique for designed seeding of individual living cells. Tissue Eng 11:1658–1666. https://doi.org/10.1089/ten.2005.11.1658
231. Okamoto T, Suzuki T, Yamamoto N (2000) Microarray fabrication with covalent attachment of DNA using Bubble Jet technology. Nat Biotechnol 18:438–441. https://doi.org/10.1038/74507
232. Cui X, Boland T, D'Lima DD, Lotz MK (2012) Thermal inkjet printing in tissue engineering and regenerative medicine. Recent Pat Drug Deliv Formul 6:149–155
233. Blaeser A, Campos DFD, Puster U et al (2016) Controlling shear stress in 3D bioprinting is a key factor to balance printing resolution and stem cell integrity. Adv Healthc Mater 5:326–333. https://doi.org/10.1002/adhm.201500677
234. Morrison NF, Harlen OG (2010) Viscoelasticity in inkjet printing. Rheol Acta 49:619–632. https://doi.org/10.1007/s00397-009-0419-z
235. Sun J, Ng JH, Fuh YH et al (2009) Comparison of micro-dispensing performance between micro-valve and piezoelectric printhead. Microsyst Technol 15:1437–1448. https://doi.org/10.1007/s00542-009-0905-3
236. Horváth L, Umehara Y, Jud C et al (2015) Engineering an in vitro air-blood barrier by 3D bioprinting. Sci Rep 5:7974. https://doi.org/10.1038/srep07974
237. Tasoglu S, Demirci U (2013) Bioprinting for stem cell research. Trends Biotechnol 31:10–19. https://doi.org/10.1016/j.tibtech.2012.10.005
238. Lee V, Singh G, Trasatti JP et al (2013) Design and fabrication of human skin by three-dimensional bioprinting. Tissue Eng C Methods 20:473–484. https://doi.org/10.1089/ten.tec.2013.0335
239. Derby B (2015) Additive manufacture of ceramics components by inkjet printing. Engineering 1:113–123. https://doi.org/10.15302/J-ENG-2015014
240. Guillotin B, Guillemot F (2011) Cell patterning technologies for organotypic tissue fabrication. Trends Biotechnol 29:183–190. https://doi.org/10.1016/j.tibtech.2010.12.008
241. Xiong R, Zhang Z, Chai W et al (2015) Freeform drop-on-demand laser printing of 3D alginate and cellular constructs. Biofabrication 7:045011. https://doi.org/10.1088/1758-5090/7/4/045011
242. Yan J, Huang Y, Chrisey DB (2012) Laser-assisted printing of alginate long tubes and annular constructs. Biofabrication 5:015002. https://doi.org/10.1088/1758-5082/5/1/015002
243. Guillemot F, Guillotin B, Fontaine A et al (2011) Laser-assisted bioprinting to deal with tissue complexity in regenerative medicine. MRS Bull 36:1015–1019. https://doi.org/10.1557/mrs.2011.272
244. Kattamis NT, Purnick PE, Weiss R, Arnold CB (2007) Thick film laser induced forward transfer for deposition of thermally and mechanically sensitive materials. Appl Phys Lett 91:171120. https://doi.org/10.1063/1.2799877
245. Schiele NR, Corr DT, Huang Y et al (2010) Laser-based direct-write techniques for cell printing. Biofabrication 2:032001. https://doi.org/10.1088/1758-5082/2/3/032001
246. Arcaute K, Mann BK, Wicker RB (2006) Stereolithography of three-dimensional bioactive poly(ethylene glycol) constructs with encapsulated cells. Ann Biomed Eng 34:1429–1441. https://doi.org/10.1007/s10439-006-9156-y

247. Chan V, Zorlutuna P, Jeong JH et al (2010) Three-dimensional photopatterning of hydrogels using stereolithography for long-term cell encapsulation. Lab Chip 10:2062–2070. https://doi.org/10.1039/C004285D
248. Jia J, Richards DJ, Pollard S et al (2014) Engineering alginate as bioink for bioprinting. Acta Biomater 10:4323–4331. https://doi.org/10.1016/j.actbio.2014.06.034
249. Bryant SJ, Nuttelman CR, Anseth KS (2000) Cytocompatibility of UV and visible light photoinitiating systems on cultured NIH/3T3 fibroblasts in vitro. J Biomater Sci Polym Ed 11:439–457. https://doi.org/10.1163/156856200743805
250. Williams CG, Malik AN, Kim TK et al (2005) Variable cytocompatibility of six cell lines with photoinitiators used for polymerizing hydrogels and cell encapsulation. Biomaterials 26:1211–1218. https://doi.org/10.1016/j.biomaterials.2004.04.024
251. Fedorovich NE, Oudshoorn MH, van Geemen D et al (2009) The effect of photopolymerization on stem cells embedded in hydrogels. Biomaterials 30:344–353. https://doi.org/10.1016/j.biomaterials.2008.09.037
252. Fairbanks BD, Schwartz MP, Bowman CN, Anseth KS (2009) Photoinitiated polymerization of PEG-diacrylate with lithium phenyl-2,4,6-trimethylbenzoylphosphinate: polymerization rate and cytocompatibility. Biomaterials 30:6702–6707. https://doi.org/10.1016/j.biomaterials.2009.08.055
253. Ma X, Qu X, Zhu W et al (2016) Deterministically patterned biomimetic human iPSC-derived hepatic model via rapid 3D bioprinting. Proc Natl Acad Sci 113:2206–2211. https://doi.org/10.1073/pnas.1524510113
254. Miri AK, Nieto D, Iglesias L et al (2018) Microfluidics-enabled multimaterial maskless stereolithographic bioprinting. Adv Mater 30:1800242. https://doi.org/10.1002/adma.201800242
255. Bertlein S, Brown G, Lim KS et al (2017) Thiol–ene clickable gelatin: a platform bioink for multiple 3D biofabrication technologies. Adv Mater 29:1703404. https://doi.org/10.1002/adma.201703404
256. Sun AX, Lin H, Beck AM et al (2015) Projection stereolithographic fabrication of human adipose stem cell-incorporated biodegradable scaffolds for cartilage tissue engineering. Front Bioeng Biotechnol 3:115. https://doi.org/10.3389/fbioe.2015.00115
257. Cheung YK, Gillette BM, Zhong M et al (2007) Direct patterning of composite biocompatible microstructures using microfluidics. Lab Chip 7:574–579. https://doi.org/10.1039/B700869D
258. Ovsianikov A, Mühleder S, Torgersen J et al (2014) Laser photofabrication of cell-containing hydrogel constructs. Langmuir 30:3787–3794. https://doi.org/10.1021/la402346z
259. Tromayer M, Dobos A, Gruber P et al (2018) A biocompatible diazosulfonate initiator for direct encapsulation of human stem cells via two-photon polymerization. Polym Chem 9:3108–3117. https://doi.org/10.1039/C8PY00278A
260. Van Hoorick J, Gruber P, Markovic M et al (2017) Cross-linkable gelatins with superior mechanical properties through carboxylic acid modification: increasing the two-photon polymerization potential. Biomacromolecules 18:3260–3272. https://doi.org/10.1021/acs.biomac.7b00905
261. Ovsianikov A, Deiwick A, Van Vlierberghe S et al (2011) Laser fabrication of three-dimensional cad scaffolds from photosensitive gelatin for applications in tissue engineering. Biomacromolecules 12:851–858. https://doi.org/10.1021/bm1015305
262. Stichler S, Böck T, Paxton N et al (2017) Double printing of hyaluronic acid/poly(glycidol) hybrid hydrogels with poly(-caprolactone) for MSC chondrogenesis. Biofabrication 9:044108. https://doi.org/10.1088/1758-5090/aa8cb7
263. Daly AC, Cunniffe GM, Sathy BN et al (2016) 3D bioprinting of developmentally inspired templates for whole bone organ engineering. Adv Healthc Mater 5:2353–2362. https://doi.org/10.1002/adhm.201600182
264. Melchels FPW, Blokzijl MM, Levato R et al (2016) Hydrogel-based reinforcement of 3D bioprinted constructs. Biofabrication 8:035004. https://doi.org/10.1088/1758-5090/8/3/035004

265. Ozbolat IT, Chen H, Yu Y (2014) Development of 'Multi-arm Bioprinter' for hybrid biofabrication of tissue engineering constructs. Robot Comput Integr Manuf 30:295–304. https://doi.org/10.1016/j.rcim.2013.10.005
266. Daly AC, Kelly DJ (2019) Biofabrication of spatially organised tissues by directing the growth of cellular spheroids within 3D printed polymeric microchambers. Biomaterials 197:194–206. https://doi.org/10.1016/j.biomaterials.2018.12.028
267. Shanjani Y, Pan CC, Elomaa L, Yang Y (2015) A novel bioprinting method and system for forming hybrid tissue engineering constructs. Biofabrication 7:045008. https://doi.org/10.1088/1758-5090/7/4/045008
268. Xu T, Binder KW, Albanna MZ et al (2012) Hybrid printing of mechanically and biologically improved constructs for cartilage tissue engineering applications. Biofabrication 5:015001. https://doi.org/10.1088/1758-5082/5/1/015001
269. Ovsianikov A, Gruene M, Pflaum M et al (2010) Laser printing of cells into 3D scaffolds. Biofabrication 2:014104. https://doi.org/10.1088/1758-5082/2/1/014104
270. Bertlein S, Hikimoto D, Hochleitner G et al (2018) Development of endothelial cell networks in 3D tissues by combination of melt electrospinning writing with cell-accumulation technology. Small 14:1701521. https://doi.org/10.1002/smll.201701521
271. Visser J, Melchels FPW, Jeon JE et al (2015) Reinforcement of hydrogels using three-dimensionally printed microfibres. Nat Commun 6:6933. https://doi.org/10.1038/ncomms7933
272. Castilho M, Hochleitner G, Wilson W et al (2018) Mechanical behavior of a soft hydrogel reinforced with three-dimensional printed microfibre scaffolds. Sci Rep 8:1245. https://doi.org/10.1038/s41598-018-19502-y
273. O'Bryan CS, Bhattacharjee T, Niemi SR et al (2017) Three-dimensional printing with sacrificial materials for soft matter manufacturing. MRS Bull 42:571–577. https://doi.org/10.1557/mrs.2017.167
274. Shi L, Carstensen H, Hoelzl K et al (2017) Dynamic coordination chemistry enables free directional printing of biopolymer hydrogel. Chem Mater 29(14):5816–5823
275. Moxon SR, Cooke ME, Cox SC et al (2017) Suspended manufacture of biological structures. Adv Mater 29:1605594. https://doi.org/10.1002/adma.201605594
276. Bhattacharjee T, Gil CJ, Marshall SL et al (2016) Liquid-like solids support cells in 3D. ACS Biomater Sci Eng 2:1787–1795. https://doi.org/10.1021/acsbiomaterials.6b00218
277. Wu W, DeConinck A, Lewis JA (2011) Omnidirectional printing of 3D microvascular networks. Adv Mater 23(24):H178–H183
278. Kolesky DB, Truby RL, Gladman AS et al (2014) 3D bioprinting of vascularized, heterogeneous cell-laden tissue constructs. Adv Mater 26:3124–3130. https://doi.org/10.1002/adma.201305506
279. Bertassoni LE, Cecconi M, Manoharan V et al (2014) Hydrogel bioprinted microchannel networks for vascularization of tissue engineering constructs. Lab Chip 14:2202–2211. https://doi.org/10.1039/C4LC00030G
280. Lee VK, Kim DY, Ngo H et al (2014) Creating perfused functional vascular channels using 3D bio-printing technology. Biomaterials 35:8092–8102. https://doi.org/10.1016/j.biomaterials.2014.05.083
281. Kim G, Ahn S, Kim Y et al (2011) Coaxial structured collagen–alginate scaffolds: fabrication, physical properties, and biomedical application for skin tissue regeneration. J Mater Chem 21:6165–6172. https://doi.org/10.1039/C0JM03452E
282. Zhang Y, Yu Y, Chen H, Ozbolat IT (2013) Characterization of printable cellular micro-fluidic channels for tissue engineering. Biofabrication 5:025004. https://doi.org/10.1088/1758-5082/5/2/025004
283. Zhang Y, Yu Y, Akkouch A et al (2014) In vitro study of directly bioprinted perfusable vasculature conduits. Biomater Sci 3:134–143. https://doi.org/10.1039/C4BM00234B
284. Gao Q, He Y, Fu J et al (2015) Coaxial nozzle-assisted 3D bioprinting with built-in microchannels for nutrients delivery. Biomaterials 61:203–215. https://doi.org/10.1016/j.biomaterials.2015.05.031

285. Jia W, Gungor-Ozkerim PS, Zhang YS et al (2016) Direct 3D bioprinting of perfusable vascular constructs using a blend bioink. Biomaterials 106:58–68. https://doi.org/10.1016/j.biomaterials.2016.07.038
286. Hardin JO, Ober TJ, Valentine AD, Lewis JA (2015) Microfluidic printheads for multimaterial 3D printing of viscoelastic inks. Adv Mater 27:3279–3284. https://doi.org/10.1002/adma.201500222
287. Foresti D, Kroll KT, Amissah R et al (2018) Acoustophoretic printing. Sci Adv 4:eaat1659. https://doi.org/10.1126/sciadv.aat1659
288. Liu W, Zhang YS, Heinrich MA et al (2017) Rapid continuous multimaterial extrusion bioprinting. Adv Mater 29:1604630. https://doi.org/10.1002/adma.201604630
289. Ober TJ, Foresti D, Lewis JA (2015) Active mixing of complex fluids at the microscale. Proc Natl Acad Sci 112:12293–12298. https://doi.org/10.1073/pnas.1509224112
290. Villar G, Graham AD, Bayley H (2013) A tissue-like printed material. Science 340:48–52
291. Graham AD, Olof SN, Burke MJ et al (2017) High-resolution patterned cellular constructs by droplet-based 3D printing. Sci Rep 7(1):7004
292. Nahmias Y, Schwartz RE, Verfaillie CM, Odde DJ (2005) Laser-guided direct writing for three-dimensional tissue engineering. Biotechnol Bioeng 92:129–136. https://doi.org/10.1002/bit.20585
293. Tsuda Y, Shimizu T, Yamato M et al (2007) Cellular control of tissue architectures using a three-dimensional tissue fabrication technique. Biomaterials 28:4939–4946. https://doi.org/10.1016/j.biomaterials.2007.08.002
294. Ong CS, Fukunishi T, Zhang H et al (2017) Biomaterial-free three-dimensional bioprinting of cardiac tissue using human induced pluripotent stem cell derived cardiomyocytes. Sci Rep 7:4566. https://doi.org/10.1038/s41598-017-05018-4
295. Yamato M, Okano T (2004) Cell sheet engineering. Mater Today 7:42–47. https://doi.org/10.1016/S1369-7021(04)00234-2
296. Hannachi IE, Yamato M, Okano T (2009) Cell sheet technology and cell patterning for biofabrication. Biofabrication 1:022002. https://doi.org/10.1088/1758-5082/1/2/022002
297. Bakirci E, Toprakhisar B, Zeybek M et al (2017) Cell sheet based bionk for 3D bioprinting applications. Biofabrication 9(2):024105
298. Nishiguchi A, Yoshida H, Matsusaki M, Akashi M (2011) Rapid construction of three-dimensional multilayered tissues with endothelial tube networks by the cell-accumulation technique. Adv Mater 23:3506–3510. https://doi.org/10.1002/adma.201101787
299. Todhunter ME, Jee NY, Hughes AJ et al (2015) Programmed synthesis of three-dimensional tissues. Nat Methods 12:975–981. https://doi.org/10.1038/nmeth.3553
300. Blakely AM, Manning KL, Tripathi A, Morgan JR (2014) Bio-pick, place, and perfuse: a new instrument for three-dimensional tissue engineering. Tissue Eng C Methods 21:737–746. https://doi.org/10.1089/ten.tec.2014.0439
301. Ovsianikov A, Khademhosseini A, Mironov V (2018) The synergy of scaffold-based and scaffold-free tissue engineering strategies. Trends Biotechnol 36:348–357. https://doi.org/10.1016/j.tibtech.2018.01.005
302. Caliari SR, Burdick JA (2016) A practical guide to hydrogels for cell culture. Nat Methods 13:405
303. Huh D, Torisawa Y, Hamilton GA et al (2012) Microengineered physiological biomimicry: organs-on-chips. Lab Chip 12:2156–2164. https://doi.org/10.1039/C2LC40089H
304. Nguyen DG, Funk J, Robbins JB et al (2016) Bioprinted 3D primary liver tissues allow assessment of organ-level response to clinical drug induced toxicity in vitro. PLoS One 11:e0158674. https://doi.org/10.1371/journal.pone.0158674
305. Inman JL, Robertson C, Mott JD, Bissell MJ (2015) Mammary gland development: cell fate specification, stem cells and the microenvironment. Development 142:1028–1042. https://doi.org/10.1242/dev.087643
306. Rossi G, Manfrin A, Lutolf MP (2018) Progress and potential in organoid research. Nat Rev Genet 19:671. https://doi.org/10.1038/s41576-018-0051-9

307. Kamei M, Brian Saunders W, Bayless KJ et al (2006) Endothelial tubes assemble from intracellular vacuoles *in vivo*. Nature 442:453–456. https://doi.org/10.1038/nature04923
308. Paşca SP (2019) Assembling human brain organoids. Science 363:126–127. https://doi.org/10.1126/science.aau5729
309. Binder KW, Zhao W, Aboushwareb T et al (2010) In situ bioprinting of the skin for burns. J Am Coll Surg 211:S76. https://doi.org/10.1016/j.jamcollsurg.2010.06.198
310. Skardal A, Mack D, Kapetanovic E et al (2012) Bioprinted amniotic fluid-derived stem cells accelerate healing of large skin wounds. Stem Cells Transl Med 1:792–802. https://doi.org/10.5966/sctm.2012-0088
311. Roh K-H, Nerem RM, Roy K (2016) Biomanufacturing of therapeutic cells: state of the art, current challenges, and future perspectives. Annu Rev Chem Biomol Eng 7:455–478. https://doi.org/10.1146/annurev-chembioeng-080615-033559
312. Liaw C-Y, Guvendiren M (2017) Current and emerging applications of 3D printing in medicine. Biofabrication 9:024102. https://doi.org/10.1088/1758-5090/aa7279

Chapter 2
Materials as Bioinks and Bioink Design

Paula Camacho, Hafiz Busari, Kelly B. Seims, John W. Tolbert, and Lesley W. Chow

Abstract This chapter summarizes the major concepts and recent progress in the design and formulation of bioinks for 3D bioprinting. Bioinks encompass cells and materials designed for processing by an automated biofabrication technique, such as direct-write, inkjet, stereolithography (SLA), or laser-induced forward transfer (LIFT) technologies, with each having its own requirements for material properties to fabricate specific tissue constructs. There are two major types of bioinks: (1) scaffold-free, consisting of cellular aggregates, and (2) scaffold-based, comprised of biomaterials with encapsulated cells. These bioinks can be composed of single materials or blends of multiple components to develop constructs tailored to preferred printing techniques and applications. Key parameters important in material selection include printability, mechanical properties, degradation, biochemical functionality, cell viability, and biocompatibility. Single-component hydrogels have limitations since properties that enhance cell viability and function often contrast with those that facilitate printing of mechanically robust constructs. More complex formulations, such as multi-material bioinks, interpenetrating networks, and nanocomposite bioinks, expand the range of properties and techniques that can be achieved for desired applications. Future directions will demonstrate how bioinks can be optimized and exploited to engineer native-like tissue constructs with

P. Camacho
Department of Bioengineering, Lehigh University, Bethlehem, PA, USA
e-mail: pac417@lehigh.edu

H. Busari · K. B. Seims
Department of Materials Science and Engineering, Lehigh University, Bethlehem, PA, USA
e-mail: hab216@lehigh.edu; kbs218@lehigh.edu

J. W. Tolbert
Department of Polymer Science and Engineering, Lehigh University, Bethlehem, PA, USA
e-mail: jwt318@lehigh.edu

L. W. Chow (✉)
Department of Bioengineering, Lehigh University, Bethlehem, PA, USA

Department of Materials Science and Engineering, Lehigh University, Bethlehem, PA, USA
e-mail: lac415@lehigh.edu

M. Guvendiren (ed.), *3D Bioprinting in Medicine*,
https://doi.org/10.1007/978-3-030-23906-0_2

spatially and temporally organized biochemical and biophysical cues and tissue-specific cell types.

Keywords Bioink design · Polymeric materials · Hydrogels · Cellular aggregates

2.1 Introduction

A critical step in the bioprinting process involves the selection and design of ideal bioinks. Bioinks are generally defined as "*a formulation of cells suitable for processing by an automated biofabrication technology that may also contain biologically active contents and biomaterials*" [1]. Ideal bioinks must provide key characteristics, such as biocompatibility, viscoelasticity, high mechanical integrity, appropriate degradability, nontoxicity, and non-immunogenicity, to enable successful fabrication of complex biomimetic tissue constructs. To achieve this, materials used in the formulation of bioinks should be selected based on mechanical, rheological, chemical, and biological properties that support both cell viability and printability for a given application.

There are two major types of bioinks used in 3D bioprinting: (1) scaffold-free and (2) scaffold-based. Scaffold-free bioinks involve directly printing living cells into 3D constructs [2–4]. In this case, cells are first formed into clusters or aggregates (most often spheroids) and engineered for bioprinting processes. The resulting cellular aggregates are then deposited in specific patterns where they fuse and mature into larger scale functional tissues [5]. In scaffold-based bioinks, cells are embedded in polymeric hydrogels then printed into scaffolds with tissue-specific 3D architectures [6]. The hydrogel scaffold degrades, and the encapsulated cells proliferate and mature into larger-scale tissues. Scaffold-based bioink strategies are advantageous for generating biomimetic tissues but have intrinsic limitations, such as restricted cell proliferation, migration, and colonization since the cells are immobilized within the material. Scaffold-free bioinks, on the other hand, better facilitate cellular interactions, including homo-cellular and hetero-cellular interactions that enable the generation of tissue with close biomimicry and cell functionality for longer time periods [7]. However, a very high number of cells are needed to prepare a sufficient number of cellular aggregates for bioprinting. These numbers can reach a few hundred million cells depending on cell size and how quickly the cells deposit extracellular matrix (ECM). In general, expanding cell populations to these numbers is labor-intensive and costly, and some cell types cannot proliferate quickly, limiting their applicability and availability [8]. While both approaches have their advantages and disadvantages, this chapter mainly focuses on materials used in the design and formulation of scaffold-based bioinks as they are the most common type of bioinks used in 3D bioprinting processes. A brief overview of scaffold-free bioinks is included in *Bioink Formulations*, but more detailed discussions can be found in recent literature [3, 9].

There are a wide variety of materials used in the design and formulation of bioinks, including alginate [10], collagen [11], fibrin [12, 13], cellulose [14], silk [15,

16], and ECM-derived components [17, 18]. However, the candidacy of materials in a bioink is highly dependent on the specific tissue of interest and the 3D printing technique being utilized (Fig. 2.1). For example, in direct-write 3D printing, high viscosity cell solutions are extruded from an ink deposition nozzle, which can result in mechanical forces and shear stresses that may damage cells [19]. Therefore, bioinks exhibiting viscoelastic behavior, such as shear-thinning properties, are needed to compensate for the high shear stresses associated with the bioink's high viscosity. In inkjet printing, polymeric solutions, colloidal suspensions, and cell suspensions with relatively low viscosities are deposited at relatively high shear rates in the form of droplets that generate biologically viable constructs [20]. Bioinks used in inkjet printing should have a non-fibrous nature, so they can flow through the nozzle without clogging and exhibit rheopectic behavior. Rheopectic materials have time-dependent non-Newtonian behavior, resulting in increased viscosity as shear is applied. This triggers droplet formation due to an increase in viscosity following ejection [21]. In stereolithography (SLA), an ultraviolet laser is utilized to photopolymerize the surface of a bath of liquid polymer adhered to a build platform that supports 3D-printed constructs as they are being fabricated. Bioinks used for SLA bioprinting should be able to undergo photopolymerization as well as possess sufficient adhesion and low surface tension characteristics, so that the bioink can uniformly spread on the build platform and adhere to it without dripping [22]. Finally, in laser-induced forward transfer (LIFT), a bioink solution is deposited onto a glass slide and coated with a laser absorption layer. A laser is directed to the laser absorption layer, creating a local pressure to eject the bioink onto the substrate [23]. Low viscosity inks are used in this process to fabricate pre-designed 3D constructs at

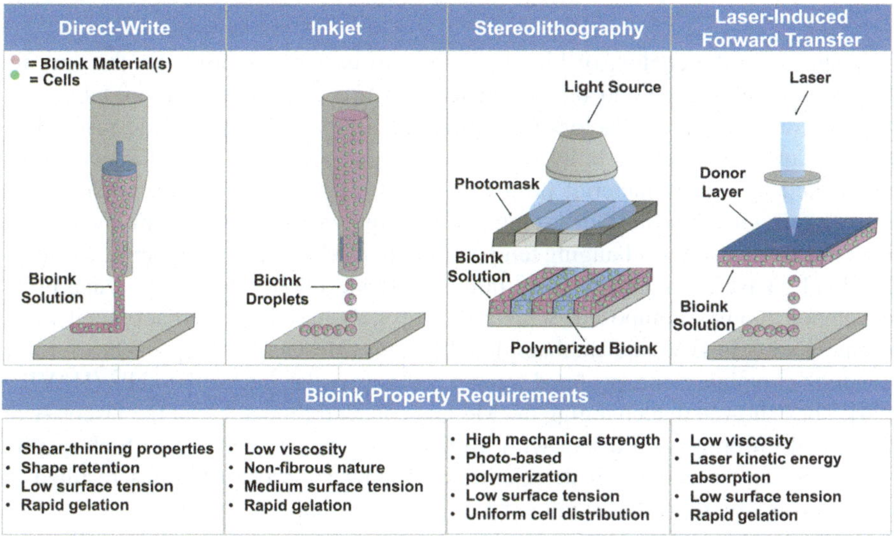

Fig. 2.1 Common bioprinting techniques and their respective bioink property requirements

high spatial resolution. The bioink must also exhibit fast cross-linking, so that the material can solidify without spreading or splashing as it is ejected. Material selection is therefore critical to ensure optimal ink properties for a desired bioprinting technique.

This chapter summarizes the major concepts and recent progress in the formulation and design of bioinks for 3D bioprinting. Specific material properties that must be considered will be discussed followed by descriptions and examples of commonly used materials. Finally, a brief discussion of state-of-the-art bioink formulations is included to provide a perspective of future directions.

2.2 Bioink Design Principles

To fabricate biomimetic structures with high print fidelity and resolution, careful control of the bioink's physicochemical and biological properties is required. Native tissues have diverse structural, mechanical, and biochemical features that are specific to each of them; therefore, it is imperative that the bioink material can be modulated to recapitulate tissue-specific properties while also providing an environment conducive to cells and tissue development. Key parameters that are important in bioink design for a desired 3D bioprinting application include printability, mechanical properties, degradation, biochemical functionality, cell viability, and biocompatibility.

2.2.1 Printability

Printability, in the context of bioinks, relates to its processability for the specific bioprinting application and the subsequent creation of accurate, high-quality biological constructs [23]. Factors that influence the printability include resolution, shape fidelity, and cell viability of the 3D-printed structure. These factors can be modulated by tuning material properties, such as viscosity, surface tension, and cross-linking mechanism. For example, tuning the viscosity of a bioink by altering material concentration, changing temperature, or modifying the number of encapsulated cells have all been shown to influence shape fidelity. Recently, Duan et al. developed bioinks composed of hybrid hydrogels based on methacrylated hyaluronic acid (Me-HA) and methacrylated gelatin (Me-Gel) to bioprint heart valve conduits containing encapsulated human aortic valvular interstitial cells (HAVIC) [24]. They showed that changing the Me-Gel concentration affected the viscosity of the bioink. Moreover, printing with a higher viscosity bioink led to high-quality structures with excellent shape fidelity while lower viscosity bioinks resulted in watery, soft structures that could not maintain their shape post-printing.

Surface tension also plays an important role when assessing the printability of a bioink. Surface tension is a result of the cohesive forces that exist between the materials present in the bioink and has been shown to influence print quality, resolution, and the dimensions of the printed structure [25]. To fabricate constructs with well-defined three-dimensional features, sufficient surface tension is needed to maintain structural integrity of the bioink as printed [26]. For example, Xu et al. showed that increasing cell concentration in alginate solutions correlated with a decrease in bio-ink surface tension during inkjet printing. The higher cell concentrations resulted in more cells being adsorbed to the interface to reduce the total free energy, causing a decrease in surface tension [27]. This also decreased the size and affected the shape of the droplet, which limits the resolution and dimensions of the printed material.

Another important characteristic to consider for printability is the extent of cross-linking within the bioink material, which often impacts the final shape and size of the printed 3D construct. Generally, materials with high cross-linking abilities are preferred in bioinks because they provide good shape fidelity after bioprinting. For example, Rutz et al. designed a bioink composed of gelatin methacrylate and multi-armed polyethylene glycol (PEG) where PEG was included to loosely cross-link gelatin methacrylate to provide the necessary viscosity for bioprinting [28]. Their strategy allowed the chemical and physical properties of the bioink to be carefully tuned. The resulting constructs showed high shape and structural fidelity without compromising cytocompatibility [28]. This introduces great potential toward developing tailorable platforms for studying cell-cell signaling and tissue morphogenesis in 3D, in addition to creating more customized, biomimetic 3D-printed tissue constructs.

2.2.2 Mechanical Properties

The mechanical properties of the selected material must enable the extruded bioink to create self-supporting structures with maintained as-printed shape fidelity and enable successful integration with previously printed layers to preserve integrity of the entire 3D-printed construct. The specific mechanical properties should also be tunable for both the specific bioprinting technique and the desired tissue application. To achieve porosity or other open architectural features, the bioink must be capable of spanning gaps of at least several 100 μm without collapsing or sagging [29]. The final printed structure may also require post-printing stabilization steps (e.g., cross-linking) to render the structure stable for in vitro culture or in vivo environments. Post-printing steps can be avoided by using multi-material bioink formulations that exploit different cross-linking motifs to enable short-term and long-term structural stability. For example, Wüst et al. combined the ionic cross-linking of alginate and the thermosensitive properties of gelatin with different concentrations of hydroxyapatite. They hypothesized that combining gelatin and alginate would improve both instantaneous stability by maintaining initial bonding between single layers due to the gelation of gelatin and long-term stability of the entire construct by cross-linking alginate with calcium. As predicted, these bioinks were able to achieve 3D printed constructs with high structural fidelity immediately after extrusion.

Furthermore, they showed that increasing hydroxyapatite concentrations, and thus increasing viscosities, improved construct stiffness and prevented uncontrolled spreading, deformation, reduced porosity, and collapse of subsequent layers [30]. Hydroxyapatite enhanced shape fidelity of the printed structure while also introducing a bioactive component for bone tissue engineering applications.

The mechanical properties of the bioink material can also influence the behavior of encapsulated cells. Stem cell differentiation, for example, can be selectively guided by changing the mechanical properties of the local environment, such as substrate stiffness [31]. Soft constructs mimic the brain while stiffer and more rigid constructs resemble muscle and bone, respectively [32]. A single mechanical property like stiffness can therefore greatly impact how cells behave and the resulting tissue they produce. Mechanical properties, such as elastic modulus, tensile strength, fracture toughness, fatigue, and elongation, of the final construct should match the desired tissue application at the time of implantation [33]. For example, Freeman and Kelly were able to tune alginate bioink stiffness to direct mesenchymal stem cell (MSC) differentiation. They used alginates with different molecular weight and systematically varied the ratio of alginate to ionic cross-linker within the bioink. They discovered that increasing alginate molecular weight increased the stiffness of the structure and promoted MSC osteogenesis (bone) over adipogenesis (fat) [34].

An often overlooked but necessary feature of cell-encapsulating inks is that they must also prevent cell sedimentation where cells settle out of the bioink, leading to inhomogeneous cell distributions. This may also lead to nozzle clogging, which disrupts the print process and reduces cell viability [35]. This has been a particular challenge for solution-based cell-encapsulating bioinks, particularly those of low viscosity. Preventing cell sedimentation is especially important when fabricating relevant-scale tissue and organ structures, which may require multiple hours of fabrication time [29]. The viscosity before printing must be optimized to ensure proper mixing and homogeneous 3D distribution of cells throughout the printing process without affecting cell viability. For example, when printing with an inkjet bioprinter for long periods of time, cell sedimentation can lead to inconsistent printing results. Parsa et al. studied the effect of adding Pluronic® as a dispersant and gentle agitation to prevent sedimentation of cells during inkjet bioprinting. The addition of Pluronic® improved the consistency of droplet formation and continuous gentle stirring reduced cell sedimentation [36].

2.2.3 Degradation

The 3D-printed construct should provide mechanical support during tissue regeneration yet be degradable in vivo to permit tissue remodeling without degrading during printing. The rate of degradation should match the rate of tissue regeneration, so that cells gradually replace the biomaterial scaffold with native ECM components [21, 23]. It is also important to note that byproducts resulting from this degradation process should not create any immunological response in the host when implanted in vivo [37]. The degradation rate depends on the specific bioink material used and should be modified according to the desired application [21]. The main mechanisms of degradation for materials commonly used as bioinks include enzymatic, hydrolytic, and ion exchange [38]. For example, collagen can be enzymatically degraded, and its type of cross-linking can influence the degradation rate [39]. Gordon et al. used type II collagen scaffolds with various cross-linking treatments, such as ultraviolet radiation and dehydrothermal treatment. Results showed that the dehydrothermal and ultraviolet light treatments produced minimal cross-linking, allowing for more rapid degradation [40]. Regarding hydrolytic degradation, Diniz et al. showed that Pluronic® has great potential for cell encapsulation and degrades rapidly, which may be desirable for certain tissue engineering applications [41]. Finally, with ion exchange, it has been shown that ionically cross-linking alginates have limited stability in vivo due to ion exchange mechanisms occurring in the body [42]. However, Kong et al. used molecular weight distribution to control the degradation of alginate hydrogels. Mixing oxidized low and high molecular weight MVGs (alginates rich in guluronic acid) at different weight fractions was shown to influence degradation rate [43]. For each material of interest, additional cross-linking may be required to increase degradation rates, particularly when mechanical support is needed during early stages of regeneration.

2.2.4 Biochemical Functionality

The bioink material should incorporate biochemical cues to direct cellular behavior, such as adhesion, migration, and differentiation [23, 37, 38]. Bioactive groups can promote tissue growth and simulate environments similar to native tissues, thus encouraging cellular function and tissue integration [44]. As discussed in other sections of this chapter, some natural polymers possess inherent bioactivity and contain cell attachment molecules, which may offer an environment more similar to that of the native ECM [38, 44]. On the other hand, synthetic polymers are more adaptable for different bioprinting applications but require chemical modification to include bioactive functional groups [21, 38, 44]. For example, the arginine-glycine-aspartic acid (RGD) peptide can be attached to the surface of biomaterials to promote cell adhesion [45]. Other molecules, such as drugs, growth factors, or other biologics, can also be incorporated for delivery to the surrounding tissue. Cross-linking chemistry or degree of cross-linking can be modified to control when these molecules are released [21]. Kundu et al. demonstrated that a photo-cross-linked hydrogel synthesized with poly(vinyl alcohol) methacrylate and silk fibroin can encapsulate growth factors and drug molecules without inhibiting their activity [46]. These same strategies can be applied for bioinks to enhance cell behavior and present or deliver specific molecules as desired. Notably, these same chemical modification techniques can be exploited to enhance post-printing construct integrity [21, 37].

2.2.5 Cell Viability

Cell viability and successful encapsulation are influenced by many factors, including the stiffness of the printed construct [23], post-processing techniques (e.g., UV photo-cross-linking), and hydrophilicity and viscosity of the bioink [37]. In 3D bioprinting, the fabrication of a layer can be described as transfer of energy (mechanical, thermal, chemical, and electromagnetic) from the bioprinting machine to the cell-laden material. This process can affect cell phenotype and viability, damage cell membranes, or alter osmotic equilibrium between cells and the external environment [47]. One of the key criteria for a well-designed bioink is to protect cells from the damaging mechanical forces experienced during extrusion [48]. The bioink not only acts as a structural medium for spatially patterning cells within a 3D structure but also serves as an incubator to support cell viability before, during, and after printing where the cells experience substantial stresses [23, 37]. These stresses may impact cell viability and health as well as their overall behavior, function, and efficacy. Stresses include the physical mixing needed to incorporate cells into the bioink matrix, shear forces encountered during extrusion through a fine diameter nozzle, and pre- or post-printing processes, such as exposure to chemicals or temperature changes often required for ink gelation [29]. It is thus vital to monitor cells before and after printing to investigate how the bioink material permits cell viability.

2.2.6 Biocompatibility

In addition to supporting cell viability, the bioink material should not induce an inflammatory response in the body [23] and should be able to be implanted in vivo without causing deleterious local or systemic reactions upon implantation and during degradation [38]. In other words, the bioink and its degradation products should be both biocompatible and cytocompatible [23]. For example, Rodriguez et al. were influenced by previous studies that showed that natural materials have high biocompatibility. They used silk-based bioinks incorporating gelatin, and their results indicated that these bioinks were highly compatible with cells introduced to the ink. The silk-gelatin-based bioprinted construct demonstrated enhanced cell viability and, after implanting in vivo, did not generate tissue inflammation in the host [49].

To ensure cytocompatibility and avoid any inflammatory response in the body, cell-encapsulating bioinks and resulting structures as well as the ink cartridge and printing substrate must maintain sterility throughout the entire fabrication process [35]. The bioink material must be compatible with proper sterilization procedures, either through sterile production or through conformity with a sterilization method before use during fabrication. Moreover, the materials should be endotoxin-free and not exceed the limits set by regulation, which may be a more critical point for biologically derived polymers like decellularized ECM than for synthetic systems [35].

Furthermore, sterility has to be assured in each step of the bioprinting procedure by an accurate control of the material reservoirs and of the building chamber [47].

2.3 Bioink Formulations

Ink development is one of the most challenging aspects in the bioprinting process. An ideal ink must satisfy biological needs for cell compatibility as well as physical and mechanical needs for the printing process. Materials must be carefully selected and tailored to produce bioinks that follow desired design principles and can be effectively bioprinted with a specific technique to build successful constructs for tissue engineering. These bioink formulations can be classified into two main categories of bioink formulations for 3D printing: cell-based bioinks for scaffold-free printing and cell-encapsulating materials to construct cell-laden scaffolds.

2.3.1 Cell-Based Bioinks and Scaffold-Free Printing

Organ printing using self-assembled tissue aggregates or spheroids is an alternative to the biodegradable scaffold-based approach in the field of tissue engineering. Cell-based bioinks use solid cellular units as building blocks to enable scaffold-free bioprinting, free of exogenous biomaterials [3]. The concept originates from developmental biology, where tissues and organs are formed without any scaffolds during embryonic development [50]. This technique enables several advantages: an automated approach for mass production of scalable and reproducible constructs, precise simultaneous 3D organization of several cell types, and creation of tissue with a high level of cell density. Furthermore, vascularization can be integrated in thick tissue constructs, and it provides a pathway for in situ application [3].

Generally, cellular aggregation techniques follow four main steps: (1) cell expansion, (2) initiation of cell aggregation, (3) cellular pellet collection, and (4) geometric molding, such as cylinder or spheroid. The bioink undergoes fully biological self-assembly without or in the presence of a temporary support layer [4, 35]. After obtaining sufficient mechanical integrity, aggregates with diameters ranging from 200 to 500 μm can be printed as building blocks to form 3D structures using inkjet technique [35, 50, 51]. However, the directly printed constructs are fragile and lack cohesive tensile strength. Therefore, successful fusion is a very important process for the formation of 3D structures that rely on the cohesive ability of multicellular aggregates and additive properties. The accumulation of ECM, associated restriction of cell motility, and enhancement of tissue cohesion in tissue spheroids can change kinetics or impede the tissue spheroid fusion process [50].

One significant advantage of this technique is accelerated tissue organization and ability to direct the formation of complex tissue structures. During cell aggregation, single- or multi-cellular aggregates fuse together through cell–cell interactions, resulting in the organization of multiple cell types in a single construct [50]. Also, tissue spheroids are thought to possess material properties that can replicate the mechanical and functional properties of the native tissue ECM. Bioprinted self-assembling cellular spheroids may produce a suitable ECM environment by themselves to form native-like complex tissues [50].

2.3.2 Cell-Encapsulating Materials and Scaffold-Based Printing

Biomaterials that qualify as a bioink must serve as a cell-delivery medium during formulation and processing, with hydrogels being the most common material class of choice [1]. Hydrogels are three-dimensional (3D) networks of hydrophilic, cross-linked polymers with a high water content that resemble the ECM in biological tissues [52, 53]. These materials generally exhibit good biocompatibility and high permeability for oxygen, nutrients, and other water-soluble metabolites, making them attractive for use in cell encapsulation [54]. They provide a highly suitable environment for embedded cells, enabling them to migrate in any direction in 3D and communicate with each other through a porous flexible network [21].

Once a desired material has been selected, additional factors with the actual 3D bioprinting process must be considered that may influence bioink formulation. For example, the cell-loaded bioink has to be stable in the reservoir for the duration of the printing procedure, typically for at least several minutes depending on the size and complexity of the structure to be printed. In addition, it must maintain the rheological properties needed for the specific bioprinting technique without clogging the nozzle and facilitate a homogeneous three-dimensional distribution of the cells [35]. The hydrogel material also needs to protect encapsulated cells from any damaging effects of flow by having an intermediate stiffness ($G' \sim 30$ Pa) and optimized viscoelastic properties [48]. To optimize 3D bioprinting, some studies have developed bioink formulations using hydrogels that gel before printing. For example, peptide-polymer hybrids that form physical cross-links create weak hydrogel networks, which protect cells during printing and prevent cell sedimentation and nozzle clogging [55]. However, the majority of materials used for bioink formulation are precursors that need to be cross-linked into hydrogels after deposition.

Hydrogel materials in tissue engineering are classified into two groups: naturally derived, such as gelatin, fibrin, collagen, chitosan, and alginate; and synthetically derived, such as Pluronic® or polyethylene glycol (PEG). Cell-encapsulating bioink materials can be composed of single materials or blends of multiple components to develop constructs tailored to desired printing techniques and applications. Table 2.1 details the key characteristics and typical printing method for natural and synthetic polymers commonly used as bioink materials. Figure 2.2 shows representative images of selected cell-laden natural and synthetic polymer scaffolds to illustrate cell viability after bioprinting. Natural polymers are advantageous because of their

Table 2.1 Natural and synthetic polymers commonly used as bioinks

Polymer	Structure and/or source	Advantages	Disadvantages	Printing technique
Agarose	Linear polysaccharide derived from seaweed	Fast gelation; readily available at low cost	Poor cell adhesion and viability	Extrusion
Alginate	Linear copolymer polysaccharide derived from brown algae	Structural diffusivity	Bioinert	Extrusion, inkjet, or laser-assisted
Cellulose	Linear polysaccharide found in plants	Environmental response; easily processable	Typically blended because of poor intrinsic properties	Extrusion
Collagen	Protein found in native ECM	Good cell adhesion	Lengthy processing time because of cross-linking requirement	Extrusion, inkjet
Decellularized ECM (dECM)	Varied structures derived from native ECM	Extremely high cell adhesion; excellent biocompatibility	Poor mechanical properties	Extrusion
Fibrin	Protein formed by enzymatic reaction between thrombin and fibrinogen	Biodegradable	Weak mechanical properties	Extrusion, inkjet
Hyaluronic acid (HA)	Glycosaminoglycan found in native ECM	Very good cell adhesion	Poor mechanical stability	Extrusion
Pluronic	Synthetic triblock copolymer	Biocompatible; water soluble	Unstable once printed	Extrusion
PEG	Synthetic linear or multi-armed polymer	Highly tunable mechanical properties	Bioinert	Inkjet, stereolithography
Silk	Protein derived from silkworms and spiders	Exceptional cell permeability; extremely high tensile strength	Lower cell viability	Extrusion

inherent biocompatibility and biodegradability [23, 60], while synthetic polymers provide advantages not found in natural polymers, such as tunable mechanical properties, photo-cross-linking ability, and pH and temperature reactivity, among others [37]. Additionally, both types of polymers can be blended together to exploit the advantageous properties of each for additional functionality and tunability.

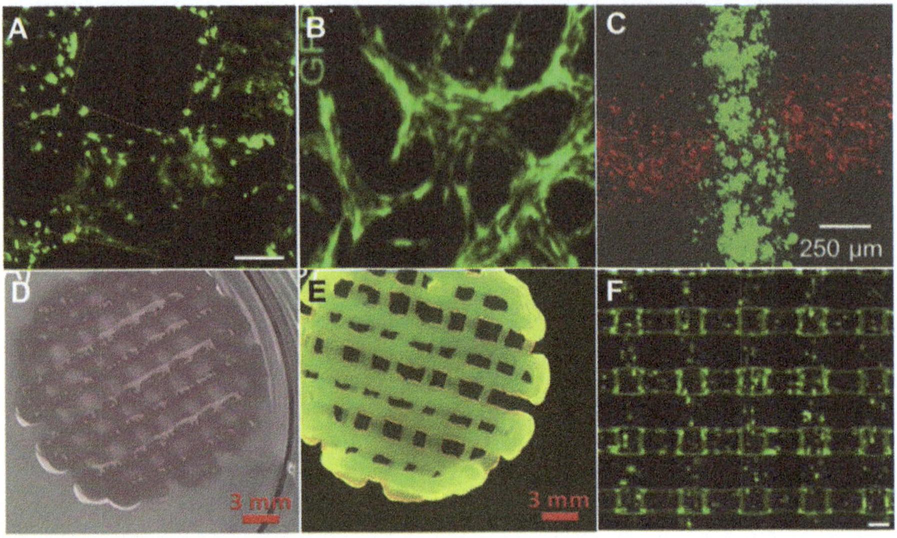

Fig. 2.2 Examples of natural and synthetic materials bioprinted into cell-laden 3D structures with high cell viability with live cells labeled with green fluorescent probe, unless otherwise noted. (**a**) Hyaluronic acid with human mesenchymal stem cells (hMSCs) (scale bar = 500 μm). Reprinted with permission from Ouyang et al. [56]. (**b**) Gelatin methacrylate (GelMA) with human umbilical vascular endothelial cells (HUVECs). Reprinted with permission from Nichol et al. [57]. (**c**) Confocal microscopy image of prelabeled human adipose-derived stem cells (hASCs) in red and fibroblasts (NIH 3T3) in green printed with MITCH-Alginate into a pattern of perpendicular lines (top–down view). Reprinted with permission from Dubbin et al. [55]. (**d**) Macroscopic and (**e**) fluorescence images of silk/poly(ethylene glycol) (PEG) with hMSCs. Reprinted with permission from Zheng et al. [58]. (**f**) Pluronic® with NIH 3T3 fibroblasts (scale bar = 200 μm). Reprinted with permission from Gioffredi et al. [59]

2.3.2.1 Natural Polymers

Natural polymers are obtained from sources found in nature and are typically plant-based. They are used for their intrinsic bioactivity, ease of formulation, biocompatibility, biodegradation, and their ability to self-assemble [23, 37]. They can also provide tissue-specific biochemical and physical stimuli to guide desirable cell behavior, including migration, proliferation, differentiation, and maturation. For use in bioprinting, natural polymers have been employed in several ways, such as by exploiting natural temperature sensitivity or ionic interactions to facilitate extrusion or covalently adding functional groups to induce chemical cross-linking [61]. The following sections detail natural polymers commonly used as bioinks.

2.3.2.1.1 Agarose

The most widely used natural polymer in bioprinting is agarose. Agarose is a polysaccharide with a linear chain obtained from seaweed and is used in a wide range of applications because of its advantageous gelation properties. Agarose forms hydrogels via hydrogen bonding rather than through covalent interactions, allowing the gelled structure to be thermally disrupted and naturally cross-link [62]. However, this gelation behavior can become a drawback during 3D printing due to the resulting high viscosity bioinks [63]. Highly viscous bioinks can be detrimental to cell viability because the high shear stresses required for printing can damage cells. To prevent cell damage, lower viscosity bioinks are used, which are inherently less mechanically stable. More resolute and tunable 3D printed structures can be constructed by creating agarose blends [64] or by chemically modifying agarose through carboxylation, a process that combines carbon monoxide with an organic compound. Shastri et al. showed that the degree of carboxylation resulted in significant changes in the elastic modulus of the printed hydrogel with almost no changes to the shear viscosity [65]. This significantly improved cell viability compared to native agarose due to low nozzle shear stress. Additionally, agarose is not intrinsically cell-adhesive and must be modified with bioactive groups or blended with other polymers, such as collagen or fibrinogen, to promote cell adhesion and biological activity [64].

2.3.2.1.2 Alginate

Alginate is a linear copolymer polysaccharide harvested from brown algae that is biocompatible but bioinert. It is composed of long chains of individual sugar residues b-D-mannuronate (M-subunits) and a-L-guluronate (G-subunits). Two G blocks of adjacent polymer chains can be cross-linked with multivalent cations (e.g., Ca^{2+} or Ba^{2+}) through interactions with the carboxylic groups in the sugars, which leads to the formation of a gel network [66]. The viscosity of alginate solutions depends on the average molecular weight (MW), molecular weight distribution, average chain subunit ratio (G to M ratio), concentration of the polymer, and the pH of the solution [67]. For example, alginate's shear-thinning behavior can be tuned by simply changing polymer concentration [68]. Additionally, Kong et al. found that the overall stiffness of the alginate gel depends on the molecular weight distribution and more specifically the M to G subunit ratio. Alginate also provides mechanical protection to cells that prevents damage caused by extensional flow during injection bioprinting procedures. Extensional flow is the phenomenon that occurs when there is an abrupt change in the flow geometry that causes a disproportionate increase in linear velocity [48]. Modifying parameters of the bioink, such as alginate concentration, temperature, and viscosity, allow the user to optimize this material for different bioprinting techniques and is therefore a promising material for cell bioprinting applications.

Modification of the alginate chains and composition influences the mechanical behavior of the material. To be used successfully for bioprinting, the alginate bioink must enable good print fidelity and maintain the printed structure under various mechanical stresses that exist either in a bioreactor or at an implant site. One common approach is to increase the concentration of high-MW alginates in the bioink to increase construct stiffness. However, this leads to an increase in viscosity of the pre-gelled solution, which is often undesirable as it increases the difficulty of processing, and cells mixed with the polymer can be damaged by high shear forces needed during mixing or injection [69]. Manipulation of the MW and MW distribution, such as specifically formulating a combination of high-MW and low-MW alginates, can overcome this problem. With this approach, the elastic modulus can be increased considerably while only minimally raising the viscosity of the pre-gel solution [70]. Another strategy is to tailor the resulting 3D-printed hydrogel mechanics by altering the degree of cross-linking and gelation time [69, 71]. Figure 2.2c shows an example from the Heilshorn group of two-component bioink that produces a shear-thinning hydrogel with two distinct cross-linking steps. This weak hydrogel prevents cell sedimentation and provides significant mechanical protection from membrane damage during printing while providing mechanical support post-printing and maintaining long-term print fidelity [55].

Alginate is an ideal material for bioprinting because of its ability to hold water and other molecules while simultaneously allowing diffusion through the structure [38]. Due to its bioinert nature, similarly to agarose, alginate is commonly modified with functional groups like RGD peptides [72] or blended with other polymers, like collagen, to promote cell adhesion and bioactivity. Once blended, alginate-based bioinks are highly advantageous because their diffusive nature creates channels that allow nutrient flow through a printed structure [37].

2.3.2.1.3 Collagen

Collagen is a protein found in the native ECM that is used because of its excellent biocompatibility properties and ability to cross-link. Collagen can be cross-linked through a change in temperature or pH or through exposure to riboflavin. Cross-linking leads to an increase in mechanical properties, such as tensile strength, viscoelasticity, and elastic modulus [73]. To achieve desired properties, collagen cross-linking, though advantageous, does require additional processing time for gelation (i.e., ~30 min at 37 °C) [73]. This limits its use for bioprinting, so collagen is often blended with other biopolymers, like agarose and alginate, to alleviate this issue.

The denatured form of collagen is gelatin, which is also used in bioinks because of its thermoresponsive gelation properties. Gelatin is used in blends with other polymers, such as alginate, because it promotes cell adhesion and increases the viscosity and viscoelastic properties of the bioink [67]. A commonly used synthetic gelatin blend is gelatin methacrylate (GelMA). GelMA is easily printable at room temperatures and can be UV photo-cross-linked to improve the mechanical properties [57]. A representative image of a GelMA scaffold printed with human umbilical vein endothelial cells (HUVECs) is shown in Fig. 2.2b [57]. In another example, Billiet et al. demonstrated that precise temperature control during the printing process of gelatin methacrylamide results in the possibility to fabricate constructs displaying an interconnected pore network in the range of 10–20 w/v%. Control over the deposited strand dimensions can be guaranteed due to the physical properties of gelatin methacrylamide hydrogels and machine-operating parameters. As a result, constructs having the desired stiffness and high shape reliability can be designed [74].

2.3.2.1.4 Fibrin

Fibrin hydrogels are made by enzymatically reacting thrombin and fibrinogen, proteins found in blood that help promote clotting. Like many other hydrogels, fibrin is highly biocompatible and biodegradable, but its mechanical properties are less advantageous. Because of its weak mechanical properties, fibrin is often blended with other polymers, such as hyaluronic acid (HA), to provide the ability to 3D print by increasing the viscosity of the ink [12]. For example, HA-fibrin hydrogels laden with both adipose stem cells (ASCs) and endothelial colony-forming cells (ECFCs) were created by Gruene et al. to demonstrate tunability of the HA-fibrin bioink. The percentage of HA was increased until the viscosity of the bioink was acceptable for bioprinting [75]. This strategy showed promise for use in vascular applications.

2.3.2.1.5 Hyaluronic Acid (HA)

Hyaluronic acid (HA), a polysaccharide found in the ECM of cartilage and connective tissues, is useful because of its cell adhesion properties. However, HA alone is not a viable polymer for printing because of its poor mechanical properties and slow gelation time and is instead chemically modified or used as a base in other polymer blends to achieve 3D printability [76]. For example, the Burdick group developed methacrylated HA (MeHA) to create photo-cross-linkable bioinks to increase the mechanical stiffness and long-term stability of the printed constructs. Increasing irradiation to UV light lead to an increase in MeHA gel rigidity [77]. They also modified HA with self-assembling motifs to improve printability and cell encapsulation, as shown in Fig. 2.2a [56]. Additionally, HA can be cross-linked with other polymers to increase the stability of the printed structure while maintaining cell adhesion and cell viability [78]. As noted above, HA has been blended with fibrin to increase bioink printability and for use in vascular applications. It can also be blended with poly(ethylene glycol) (PEG) to combine the biocompatibility and bioactivity of HA with the mechanical stability of PEG for printing a meniscus-like structure [78].

2.3.2.1.6 Cellulose (and Derivatives)

Cellulose is a naturally occurring linear polysaccharide found in plants which can be used in its original state or reacted with other chemicals to form chemical derivatives with varying properties and uses. Nanofibrillated cellulose (NFC) provides structural and mechanical support to 3D-printed constructs for forming a physiological mimetic environment. Blends with more bioactive polymers, such as cellulose/alginate or cellulose/HA, are used to provide improved cell adhesion, cell viability, and mechanical properties of the 3D-printed structures [79]. NFC is also commonly used as a shear-thinning agent for other bioinks. For example, Markstedt et al. used NFC to enhance shear-thinning properties of alginate bioink to improve its printability [80].

One common cellulose derivative is methyl-cellulose (MC), a semiflexible, linear-chain polysaccharide with a partial replacement of the hydroxyl groups with methoxy moieties [81]. Methyl-cellulose can form hydrogels, which possess a consistent fibrillar structure [81] and respond to external stimuli by changing concentration, molecular weight, salt content, and degree of methyl grafting [81, 82]. Aqueous MC can form hydrogels below 37 °C, and the addition of collagen type-I significantly improves cell adhesion and proliferation on the hydrogel [83].

2.3.2.1.7 Silk

Silk fibroin is a protein that can be harvested from silkworms and spiders. It is used in regenerative medicine and tissue engineering because of its exceptional biocompatibility, cell permeability, and tissue integration [84]. Silk fibroin hydrogels allow for cellular attachment in the absence of defined biological ligands or serum proteins and exhibit tunable stiffness by simple adjustment of the starting silk protein concentration [85]. To further improve and enhance its cell viability properties, additional biocompatible materials in silk can be included to increase the quality of the printed materials. Silk is rarely used alone but rather is blended with other polymers, such as alginate or gelatin, to offer robust mechanical properties and tailorable degradability [15, 86]. To control silk gelation and improve lubricity of the resulting hydrogels, Wang et al. developed a silk hydrogel system by blending with low molecular weight polyethylene glycol (PEG). The silk/PEG material showed tunable gelation time and mechanical properties by modifying PEG and silk concentrations [87]. Figure 2.2d, e show cell-laden silk/PEG blended constructs, illustrating a promising material bioink for mechanically stable prints [58].

2.3.2.1.8 Decellularized Extracellular Matrix (dECM)

The extracellular matrix (ECM) is the noncellular component present within all tissues and organs. It provides essential physical scaffolding for the cellular constituents and initiates crucial biochemical and biomechanical cues that are required for tissue morphogenesis, differentiation, and homeostasis. Although, fundamentally, the ECM is composed of water, proteins, and polysaccharides, each tissue has an ECM with a unique composition and topology that is generated during tissue development through a dynamic and reciprocal, biochemical, and biophysical dialogue between the various cellular components and the evolving cellular and protein microenvironment [88]. A single material cannot represent the full complexity of natural ECM or accurately recreate a microenvironment with cell–cell connections and 3D cellular organization that is typical of living tissues. Utilizing the ECM itself as a bioink is therefore of great interest for tissue engineering and regenerative medicine applications.

Decellularized extracellular matrix (dECM) is obtained by removing the cells from whole organs and tissues while maintaining the structure and composition of the ECM. The dECM is turned into a powder and dissolved into a buffer solution to create the printable material. Using 3D bioprinting, dECM bioinks can be used to construct biomimetic cell-laden scaffolds with cells spatially organized with high resolution [89]. For example, Jang et al. used tissue-specific dECM bioinks to spatially pattern different cell types in a biomimetic scaffold. The scaffold promoted vascularization, cell survival, and tissue remodeling after transplantation [90]. In another study, Pati et al. harvested multiple types of tissues to prepare dECM bioinks and printed them with a PCL framework to enhance print fidelity. This printing technique was capable of printing feature sizes as small as 100 μm. The multi-head tissue/organ building system (MtoBS) extruded cell-laden dECM bioinks at 15 °C, while the printed structures were solidified at 37 °C. To accommodate different stiffnesses of various tissues, MtoBS printed without the PCL framework for less stiff tissues or with the PCL framework for stiffer tissues or organs [18]. While there are significant advantages to using dECM bioinks, sourcing material is restricted by limited organ or tissue availability, and the decellularization process is both tedious and time-consuming.

2.3.2.2 Synthetic Polymers

Synthetic polymers are an attractive class of materials used in bioink formulations because they can be designed to mimic certain properties of natural polymers, and their mechanical properties can be fine-tuned by varying molecular weight, chemical structure, and composition to suit particular applications. However, while natural polymers provide a cell environment by mimicking native components of the ECM, most synthetic polymers do not contain the appropriate functionalities that promote cell adhesion or viability. Therefore, bioactive functional groups, such as adhesion motifs or enzymatically degradable sites, often need to be added in order to create biologically relevant structures [91]. The two most commonly used synthetic polymers in bioink formulations are Pluronic® and poly(ethylene glycol) (PEG).

2.3.2.2.1 Pluronic®

Pluronic® is a poloxamer triblock copolymer containing a hydrophilic block surrounded by two hydrophobic blocks [23, 37]. Pluronic® is used primarily for its ability to gel at room temperature and flow at 10 °C, making it ideal for extrusion-based 3D printing [23]. Pluronics allow the preparation of thermosensitive hydrogels with different properties in terms of critical gelation concentration and gelation time at physiological conditions [59]. However, Pluronic is not very stable and is used mainly as support material for printed structures [92]. Furthermore, while Pluronic® is biocompatible with cells and tissues, it does not specifically promote cell viability in long-term cell culture. Fortunately, methods to improve cell viability of Pluronic®-based bioinks have been formulated. For example, Khattak et al. showed that supplementing Pluronic® with membrane-stabilizing agents like hydrocortisone, glucose, and glycerol can dramatically increase the viability of cells encapsulated in the matrix [93]. Specifically, their results showed that adding hydrocortisone to a Pluronic® gel formulation increased the viability of human liver carcinoma cells from 2% with Pluronic® alone to 70% for up to 5 days. Müller et al. improved on this by developing a nanostructured acrylated Pluronic® network to increase the long-term biocompatibility of Pluronic® gels [94]. Through the elution of unmodified Pluronic® from a U-cross-linked network of Pluronic® and acrylated Pluronic®, they were able to increase the cell viability of encapsulated bovine chondrocytes from 62% for a pure acrylated Pluronic hydrogel to 86% for a nanostructured hydrogel for up to 14 days [94]. As seen in Fig. 2.2f, Pluronic® is a promising formulation for bioprinting due to its fast gelation in physiological conditions (a volume of 1.5 mL solution converted into a gel in 5 min), proper viscoelastic properties, and fast viscosity recovery after shearing [59].

2.3.2.2.2 Poly(Ethylene Glycol) (PEG)

PEG is a linear hydrophilic polymer that has many biomedical, industrial, and commercial uses. It is mainly used in PEG-diacrylate (PEG-DA) and PEG-methacrylate (PEG-MA) forms when creating bioinks because PEG alone does not form a hydrogel [23]. One disadvantage of using PEG-based materials for bioink formulations is the absence of bioactive moieties for cellular interactions. However, due to the robust tailorability of PEG, bioactive molecules can be tethered to mimic cell-matrix adhesions and present signaling cues. For example, short peptide sequences that mimic ECM proteins, such as fibronectin, laminin, collagen, and elastin, have been successfully attached to PEG to introduce bioactivity [45]. PEG is also advantageous because it has highly tunable mechanical properties and can be blended with other biopolymers to enhance cellular interactions. For example, Hockaday et al. printed aortic heart valves with bioinks formulated with a PEG-DA/alginate blend [95]. They saw a remarkable increase in the intrinsic elastic modulus from ~5 to ~75 kPA with the addition of PEG-DA, resulting in anatomical scaffolds exhibiting high mechanical stability and cytocompatibility. PEG is additionally useful for printing techniques like inkjet printing because it can be printed at low viscosity while still maintaining shape fidelity [96].

2.3.2.3 Multi-Material Bioinks

Single-component hydrogels, including those ones above, are limited since properties that enhance cell viability and function are often at odds with those that facilitate printing. These hydrogels can be optimized for bioprinting by increasing polymer concentration and/or cross-linking density to improve print fidelity, but these modifications can also be detrimental to encapsulated cells. For example, increasing polymer concentration or cross-linking reduces porosity and thus prevents cell spreading and migration as well as limits nutrient and oxygen diffusion [38]. Furthermore, cells thrive on porous networks with cell-binding domains to facilitate cell spreading. The presence of proteolytic cleavage sites, which are typically found in natural polymers, also permit cell migration. However, hydrogels composed of natural materials generally have weak mechanical properties that are not optimal for bioprinting while their synthetic counterparts lack bioactivity such as functional groups that promote cell adhesion or migration [54]. More complex formulations, such as multi-material bioinks, interpenetrating networks, and nanocomposite bioinks, as shown in Fig. 2.3, expand the range of properties and techniques that can be achieved for desired applications.

When bioprinting cell-laden hydrogel mixtures, two major disadvantages can be described: first, the loss of cell viability due to the printing process (discussed in previous sections); and second, obtaining adequate print fidelity that yields mechanically stable constructs without internal pore collapse and uniform distribution of cells. Multi-material bioinks provide strategies to prevent sedimentation of cells, such as the use of thickening agents like poly(ethylene glycol) diacrylate (PEGDA), for solution-phase inks and the use of gel-phase inks, such as gelatin methacrylate (GelMA) [100]. However, good print fidelity remains the main challenge, even in the absence of cells. Most bioink materials and printing methods do not result in manufactured scaffolds in which pores are fully interconnected. Conventional scaffolds have increased porosity, even though the mechanical properties of the scaffold weaken as porosity increases. In recent years, several approaches have been developed to obtain stable 3D-printed constructs without internal pore collapse.

One approach is to use sacrificial materials as support to improve the stability and fidelity of printed structures. For example, Shim et al. developed a multi-head deposition system (MHDS) to fabricate novel hybrid scaffolds that infuse HA, gelatin, and atelocollagen into a 3D scaffold consisting of synthetic polymers polycaprolactone (PCL) and poly (lactic-*co*-glycolic acid) (PLGA). This multi-material strategy used PCL and PLGA to support the overall structure while integrating bioactive materials to improve cellular interaction [101]. Using a similar concept, Kesti et al. developed a bioink from blending a thermoresponsive polymer and a photo-cross-linkable biopolymer. The blend showed rapid gelation upon contact with a

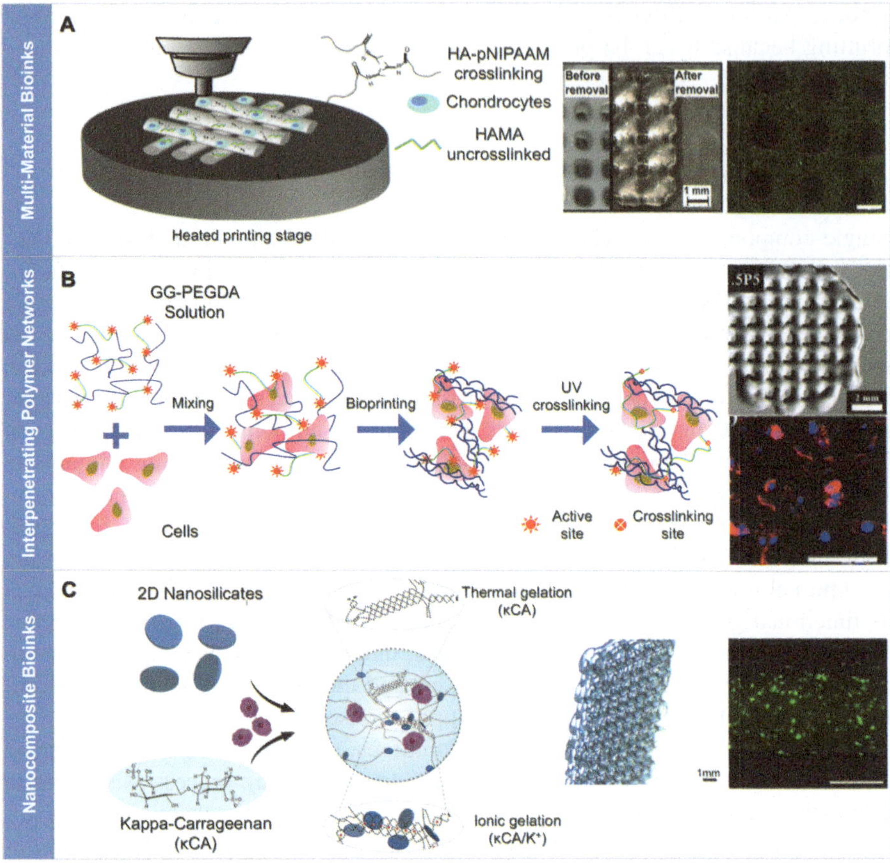

Fig. 2.3 Examples of multi-component bioink formulations. (**a**) Multi-material bioinks prepared by blending thermoresponsive polymer poly(*N*-isopropylacrylamide) grafted hyaluronic acid (HA-pNIPAAM) with methacrylated hyaluronic acid (HAMA) for high-resolution prints with good structural fidelity. Printed constructs also showed excellent cell viability with encapsulated bovine chondrocytes (scale bar = 100 μm). Reprinted with permission from Kesti et al. [97]. (**b**) Interpenetrating polymer network (IPN) bioink combining the superior shear-thinning and recovery properties of gellan gum (GG) with rapid UV cross-linking capability of poly(ethylene glycol) diacrylate (PEGDA) formulated for extrusion-based bioprinting. Excellent rheological properties of the IPN bioink enabled the printed constructs to retain shape stably after deposition without additional support. Furthermore, murine bone marrow stromal cells (BMSCs) and MC3T3-E1 mouse osteoblastic cells encapsulated in this IPN bioink exhibited high cell viability percentages over 87% during extended 3D culture. Reprinted with permission from Wu et al. [98]. (**c**) Thermo-responsive and shear-thinning nanocomposite bioinks composed by adding nanosilicates to cross-linked kappa-carrageenan (κCA) hydrogel materials. Nanosilicate-stabilized κCA networks improved the mechanical stability of 3D-printed anatomical-size structures and showed excellent biocompatibility with encapsulated MC3T3-E1 mouse preosteoclasts (scale bar = 100 μm). Reprinted with permission from Wilson et al. [99]

37 °C heated substrate, giving the printed construct its immediate structural fidelity while the secondary chemical cross-linking component provided long-term mechanical stiffness. The thermoresponsive polymer acted as temporary support for printability then was later washed away [97]. Recently, the Feinberg Lab established a 3D bioprinting technique termed Freeform Reversible Embedding of Suspended Hydrogels, or FRESH. FRESH uses a thermoreversible support bath to enable deposition of biologically relevant hydrogel inks, including alginate, fibrin, collagen type I, and Matrigel into complex, 3D biological structures. The key innovation in FRESH is the deposition and embedding of the printed hydrogel(s) within a secondary hydrogel support bath that maintains the intended structure during the print process. This significantly improves print fidelity while also in a sterile, aqueous, buffered environment compatible with cells. Once the entire 3D structure is printed, the temperature is raised to a cell-friendly 37 °C, causing the gelatin support bath to melt in a non-destructive manner [102]. Granular hydrogels are also used as support when 3D printing complex structures. The constructs are made by carefully tracing a series of programmed paths within a granular gel, using a fine hollow tip that fluidizes the medium at the point of injection of the desired material. The rapid solidification of the granular gel then traps and holds the injected material behind the moving tip. Holding material within the jammed medium allows manufacturing of finely detailed and delicate materials with nearly limitless aspect ratios and is compatible with a wide variety of materials, including silicones, hydrogels, colloids, and living cells [103].

A second strategy to improve print fidelity is blending gelatins or polymer cross-linkers with the core materials. For example, Chung et al. demonstrated that the viscosity of an alginate solution, and thereby printability, can be controlled by incorporating gelatin and modulating the mixing temperature during printing to form a gel that retains biological aspects of the original alginate solution while satisfying physical extrusion criteria [67]. Furthermore, incorporating PEG cross-linkers to loosely connect the gelatin backbone provides the necessary viscosity for bioprinting with low gelatin concentration. This strategy allows for polymer type and concentration, degree of cross-linking, as well as post-printing cross-linking to be tailored with ease to tune material properties of the bioinks and 3D-printed structures [28].

2.3.2.3.1 Interpenetrating Polymer Networks (IPNs)

Interpenetrating polymer networks (IPNs) are a strategy for strengthening hydrogels by combining two or more networks that are at least partially interlaced on a molecular scale but not covalently bonded to each other and cannot be separated unless chemical bonds are broken [104]. These materials have been shown to have increased fracture strength and toughness relative to single component networks of either of its constituent polymers [105, 106] and are widely used in a variety of biomedical applications including tissue engineering [107, 108], drug delivery [109, 110], and pharmaceutics [111]. The improved strength of IPNs come from the physical entanglement of an ionically cross-linked polymer network that serves to dissipate energy from an applied load with a covalently cross-linked polymer network that maintains the elasticity of the hydrogel. These properties are important when considering IPNs for bioink formulations because they can be used to produce bioinks with increased elasticity and stiffness.

Recently, Wu et al. reported on an IPN hydrogel formulated for extrusion 3D bioprinting with cells that combines shear-thinning and recovery properties of gellan gum with the rapid cross-linking capability of poly(ethylene glycol) diacrylate (PEGDA) [98]. As a bioink, gellan gum has many advantages over other hydrogels, including shear-thinning behavior and high gelling efficiency at physiological temperature [112]. However, the intrinsic brittleness of gellan gum restricts its printability and structural stability when manipulated. To overcome this issue, Wu et al., combined gellan gum with PEGDA to form a high strength and stretchable IPN structure. They showed the rheological properties of the gellan gum/PEGDA IPN bioink enabled the printed constructs to retain the shape after deposition without additional support. Furthermore, cells encapsulated in this hydrogel exhibited 87% viability during a long-term 3D culture of 21 days [112].

Semi-IPNs have also been explored as potential bioinks for 3D printing. In a semi-IPN, only one polymer is present as a network, which enhances miscibility in comparison to a full IPN [113]. In a study by Pescosolido et al., a semi-IPN hydrogel based on hyaluronic acid and a dextran derivate was used to form 3D bioprinted constructs [114]. They showed that the semi-IPNs exhibited suitable rheological properties for bioprinting as well as the fabrication of stable constructs with mechanical properties matching the mechanical strengths of natural tissues.

These studies show that IPNs and semi-IPNs possess features that are attractive in various bioprinting applications. However, this research area is still being developed, and more investigation is needed to elucidate the unique potential that bioinks formulated with IPNs can introduce to 3D bioprinting.

2.3.2.3.2 Nanocomposite Bioinks

Nanocomposite bioinks are formed by combining polymers with inorganic or organic nanofillers, such as carbon nanotubes, silicates, metals, metal oxides, and ceramics. These fillers consist of different mechanical, chemical, thermal, and electrical properties that can improve the structural and functional properties of polymeric materials used in desired applications. For example, silver nanoparticles have been used to improve the mechanical properties [115], bioactivity [116], and biocompatibility [117] of nanocomposite materials compared to the polymer matrix alone. Although nanocomposite bioinks with tailored properties for tissue engineering application have been developed, the enormous potential of nanocomposites in 3D bioprinting applications is yet to be fully realized.

Recent studies have shown that the addition of nanoparticles in bioink formulations can result in changes in physical and chemical characteristics, such as increased stiffness, shear-thinning characteristics, and resistance to degradation under physiological conditions. For example, Lee et al. reported the use of silica nanoparticles to improve the mechanical properties, printability, and print fidelity of a polymeric bioink mixture containing alginate and gellan gum [99]. Reversible electrostatic interactions between the cationic-modified silica nanoparticles and the anionic polymer mixture of alginate and gellan gum led to an increase in shear-thinning properties as well as an increase in the storage modulus. This led to a high-quality 3D printed structure, whereas the use of the polymeric ink without the nanoparticles led to collapse of the structure during printing. Furthermore, they were able to show that shrinkage and swelling of the printed constructs during cross-linking were significantly suppressed by the addition of nanoparticles compared to the ink without nanoparticles, resulting in high printing fidelity after cross-linking. In a similar approach, Wilson et al. reported the use of nanosilicates within kappa-carrageenan (κCA) hydrogels to fabricate bioinks with improved shear-thinning properties and gelation characteristics [99]. By tuning the kCA-nanosilicate ratio, the thermoreversible gelation could be achieved to improve printability and shape retention characteristics compared to κCA without nanosilicates. Due to the high shear-thinning characteristics of the bioink, high cell viability was observed in the printed cells.

Zhu et al. explored the incorporation of gold nanorods (GNRs) in gelatin methacryloyl and alginate pre-polymer solutions to create a nanocomposite bioink for bioprinting functional cardiac constructs [118]. Gold nanomaterials are particularly attractive in biomedical research due to their biocompatibility, versatility, and ease of modification [119]. In this study, it was shown that the incorporation of GNRs led to an amplified shear-thinning effect and improved overall printability. GNRs also introduced an electrically conductive component that promoted electrical propagation between cardiac cells necessary for a functional cardiac tissue construct. Additionally, it was shown that cardiac cells showed improved cell adhesion and organization in the printed GNR constructs when compared to constructs without the GNRs.

The development of novel nanocomposite bioinks may add unprecedented functionalities to 3D bioprinted constructs and widen the portfolio of strategies that can be explored in the 3D printing process.

2.4 Future Perspectives

2.4.1 Supramolecular Materials

Hydrogels for tissue engineering applications should be mechanically tough and capable of surviving repeated mechanical deformation. When subjected to repeated stress, bonds in conventional hydrogels can break, resulting in progressive loss of mechanical integrity. To overcome this drawback, supramolecular bioinks are currently under investigation [120, 121]. Supramolecular polymers are composed of short repeating units with functional groups that can interact non-covalently with other functional units, forming large, polymer-like entanglements. Under high stress, these non-covalent bonds are reversibly broken to dissipate energy. The reversibility of these bonds also leads to shear-thinning properties that facilitate their use in bioprinting [38].

Self-assembly is a bioinspired strategy to form physical hydrogels with shear-thinning and self-healing (time-dependent recovery) properties for injectable and modular systems [122, 123]. For example, peptides with propensity to form β-sheets can be grafted onto polymers, such as *N*-(2-hydroxypropyl)methacrylamide (HPMA) [124], poly(γ-glutamic acid) [122], or HA [125], to induce self-assembly into hybrid hydrogels. Hydrogels can also be formed by modifying polymers with host–guest chemistries, which involves host macrocyclic molecules with cavities to accept a guest molecule with complementary size and shape to generate cross-linking. In a recent study, Highley et al. developed a hydrogel-based 3D printing approach based on modified hyaluronic acid (HA) that permits the printing of shear-thinning hydrogel inks directly into self-healing support hydrogels [126]. Both materials contained guest–host chemistries for supramolecular assembly of the hydrogel. The use of a guest–host hydrogel as a support matrix deforms to accommodate the extruded material and self-heals to maintain material localization [126]. DNA hybridization represents another approach to fabricating supramolecular hydrogels. Li et al. developed supramolecular polypeptide–DNA hydrogel for rapid in situ 3D bioprinting by designing two bioinks—one containing a polypeptide–DNA conjugate and the other containing the complementary DNA linker. By alternative deposition of the components in the programmed position, designed 3D structures containing viable and functional living cells could be constructed. The resultant hydrogel combines favorable properties of both the polypeptide and DNA components, that is, it is responsive to proteases and nucleases, leading to full biodegradability and programmability of the hydrogel networks under physiological conditions [127]. Complementary peptide binding can also be used to create supramolecular hydrogels. Dubbin et al. developed a platform gel phase ink, called Recombinant-protein Alginate Platform for Injectable Dual-cross-linked ink, or RAPID, which consists of two components that undergo an initial cross-linking mechanism that exploits reversible, hetero-assembly of complementary peptide-binding domains tethered to the second component, an alginate biopolymer to improve cell viability during printing [128].

These adaptable hydrogels formed by reversible cross-links allow local modifications without requiring degradation or irreversible changes to the overall integrity [129]. Bonds formed by non-covalent interactions, such as β-sheets or host–guest pairs, can be broken and reformed, providing strategies to induce spatiotemporal changes to hydrogel composition and organization. However, the highly selective and directional nature of host–guest interactions limits the co-assembly of multiple components to those with complementary pairs, which restricts how the composition can be modified. Non-discriminant interactions, such as β-sheet formation, therefore enable temporal changes to the hydrogel composition [130].

Supramolecular hydrogels are a promising material for bioprinting due to their favorable properties for printing, such as shear-thinning and self-healing behavior, as well as their adaptable properties that allow for dynamic tissue formation.

2.4.2 Microgels

Nanogels and microgels are cross-linked spherical hydrogel particles that have nanoscale (typically 20–250 nm) and microscale (typically 1–350 mm) dimensions, respectively. They have excellent biocompatibility, high water content, tunable sizes, large surface area for multivalent bioconjugation, and abundant space to accommodate bioactive materials, such as drugs and live cells [131]. This emerging technology enables spatial deposition of different microgels containing specific cells to create organized scaffolds for native-like tissue regeneration. Cell-laden microgels with diameters between 100 and 400 μm ensure that the encapsulated cells stay safely within oxygen diffusion constraints. Additionally, important factors that influence the behavior of cells in the microenvironment, such as ECM components, soluble biomolecular signals, and biophysical cues, can be incorporated into the microgels to provide the cells with more physiologically relevant microenvironments [132].

For example, the Segura group used microgels to create microporous annealed particle (MAP) gels, which circumvent the need for material degradation before tissue ingrowth by providing a stably linked interconnected network of micropores for cell migration and bulk integration with surrounding tissues. Lattices of microgel building blocks are annealed to one another via surface functionalities to form an interconnected microporous scaffold either with or without cells present in the interconnected pores [133]. Recently, the Burdick Lab developed a granular hydrogel system, based on the interactions of modular microgel components via guest–host interactions. They fabricated cross-linked hyaluronic acid (HA) microgels with modular intraparticle covalent cross-linking that were formed into an injectable granular hydrogel using cyclodextrin and adamantane guest–host interparticle cross-linking. This hydrogel displayed shear-thinning and self-healing properties and removed the need for controlled working/cure times that are often associated with other cross-linking mechanisms [134]. These examples point to new directions and strategies for 3D bioprinting that will enable fabrication of larger and more complex engineered tissue constructs.

2.5 Conclusion

Selection and design of ideal bioinks is a critical step in 3D bioprinting. Bioinks are a formulation of materials and cells designed for processing by an automated biofabrication technique. They can be 3D-printed using direct-write, inkjet, SLA, or

LIFT technologies, each having their own requirements for material properties to fabricate accurate biomimetic and mechanically stable constructs. There are two major types of bioinks used in 3D bioprinting: (1) scaffold-free consisting of cellular aggregates and (2) scaffold-based that contain biologically active components and biomaterials for cell encapsulation. Scaffold-based bioinks involve two main classes of materials: natural and synthetic polymers. These bioinks can be composed of single materials or blends of multiple components to develop constructs tailored to desired printing techniques and applications. Key parameters important in material selection include printability, mechanical properties, degradation, biochemical functionality, cell viability, and biocompatibility. Single-component hydrogels have limitations since the properties that enhance cell viability and function often contrast with those that facilitate printing. More complex formulations, such as multi-material bioinks, interpenetrating networks, and nanocomposite bioinks, expand the range of properties and techniques that can be achieved for desired applications. As the field continues to evolve, more sophisticated materials, such as supramolecular materials and microgels, will emerge as popular bioink materials that more closely mimic the dynamic properties of native cellular microenvironments. Future directions will demonstrate how bioinks can be optimized and exploited to engineer native-like tissue constructs with spatially and temporally organized biochemical and biophysical cues and tissue-specific cell types.

Acknowledgments This work was supported by startup funds provided by Lehigh University. KBS also acknowledges the support through a President's Scholarship from Lehigh University.

References

1. Groll J, Burdick JA, Cho D, Derby B, Gelinsky M, Heilshorn SC, Jüngst T, Malda J, Mironov VA, Nakayama K, Ovsianikov A, Sun W, Takeuchi S, Yoo JJ, Woodfield TBF (2018) A definition of bioinks and their distinction from biomaterial inks. Biofabrication 11:013001. https://doi.org/10.1088/1758-5090/aaec52
2. Achilli T-M, Meyer J, Morgan JR (2012) Advances in the formation, use and understanding of multi-cellular spheroids. Expert Opin Biol Ther 12:1347–1360. https://doi.org/10.1517/14712598.2012.707181
3. Mironov V, Visconti RP, Kasyanov V, Forgacs G, Drake CJ, Markwald RR (2009) Organ printing: tissue spheroids as building blocks. Biomaterials 30:2164–2174. https://doi.org/10.1016/j.biomaterials.2008.12.084
4. Cui X, Hartanto Y, Zhang H (2017) Advances in multicellular spheroids formation. J R Soc Interface 14:20160877. https://doi.org/10.1098/rsif.2016.0877
5. Rezende RA, Pereira FDAS, Kasyanov V, Kemmoku DT, Maia I, Da Silva JVL, Mironov V (2013) Scalable biofabrication of tissue spheroids for organ printing. Proc CIRP 5:276–281. doi: https://doi.org/10.1016/j.procir.2013.01.054
6. Liu W, Heinrich MA, Zhou Y, Akpek A, Hu N, Liu X, Guan X, Zhong Z, Jin X, Khademhosseini A, Zhang YS (2017) Extrusion bioprinting of shear-thinning gelatin methacryloyl bioinks. Adv Healthc Mater 6:1–11. https://doi.org/10.1002/adhm.201601451

7. Yipeng J, Yongde X, Yuanyi W, Jilei S, Jiaxiang G, Jiangping G, Yong Y (2017) Microtissues enhance smooth muscle differentiation and cell viability of hADSCs for three dimensional bioprinting. Front Physiol 8:1–10. https://doi.org/10.3389/fphys.2017.00534
8. Ozbolat IT (2015) Scaffold-based or scaffold-free bioprinting: competing or complementing approaches? J Nanotechnol Eng Med 6:024701. https://doi.org/10.1115/1.4030414
9. Moldovan NI (2018) Progress in scaffold-free bioprinting for cardiovascular medicine. J Cell Mol Med 22:2964–2969. https://doi.org/10.1111/jcmm.13598
10. Axpe E, Oyen M (2016) Applications of alginate-based bioinks in 3D bioprinting. Int J Mol Sci 17:1976. https://doi.org/10.3390/ijms17121976
11. Yang X, Lu Z, Wu H, Li W, Zheng L, Zhao J (2018) Collagen-alginate as bioink for three-dimensional (3D) cell printing based cartilage tissue engineering. Mater Sci Eng C 83:195–201. https://doi.org/10.1016/j.msec.2017.09.002
12. England S, Rajaram A, Schreyer DJ, Chen X (2017) Bioprinted fibrin-factor XIII-hyaluronate hydrogel scaffolds with encapsulated Schwann cells and their in vitro characterization for use in nerve regeneration. Bioprinting 5:1–9. https://doi.org/10.1016/j.bprint.2016.12.001
13. Zhang K, Fu Q, Yoo J, Chen X, Chandra P, Mo X, Song L, Atala A, Zhao W (2017) 3D bioprinting of urethra with PCL/PLCL blend and dual autologous cells in fibrin hydrogel: an in vitro evaluation of biomimetic mechanical property and cell growth environment. Acta Biomater 50:154–164. https://doi.org/10.1016/j.actbio.2016.12.008
14. Sultan S, Siqueira G, Zimmermann T, Mathew AP (2017) 3D printing of nano-cellulosic biomaterials for medical applications. Curr Opin Biomed Eng 2:29–34. https://doi.org/10.1016/j.cobme.2017.06.002
15. Das S, Pati F, Choi YJ, Rijal G, Shim JH, Kim SW, Ray AR, Cho DW, Ghosh S (2015) Bioprintable, cell-laden silk fibroin-gelatin hydrogel supporting multilineage differentiation of stem cells for fabrication of three-dimensional tissue constructs. Acta Biomater 11:233–246. https://doi.org/10.1016/j.actbio.2014.09.023
16. Jose RR, Rodriguez MJ, Dixon TA, Omenetto F, Kaplan DL (2016) Evolution of bioinks and additive manufacturing technologies for 3D bioprinting. ACS Biomater Sci Eng 2:1662–1678. https://doi.org/10.1021/acsbiomaterials.6b00088
17. Shin M, Galarraga JH, Kwon MY, Lee H, Burdick JA (2018) Gallol-derived ECM-mimetic adhesive bioinks exhibiting temporal shear-thinning and stabilization behavior. Acta Biomater. pii: S1742-7061(18)30627-5. https://doi.org/10.1016/J.ACTBIO.2018.10.028
18. Pati F, Jang J, Ha DH, Won Kim S, Rhie JW, Shim JH, Kim DH, Cho DW (2014) Printing three-dimensional tissue analogues with decellularized extracellular matrix bioink. Nat Commun 5:1–11. https://doi.org/10.1038/ncomms4935
19. Donderwinkel I, van Hest JCM, Cameron NR (2017) Bio-inks for 3D bioprinting: recent advances and future prospects. Polym Chem 8:4451–4471. doi: https://doi.org/10.1039/C7PY00826K
20. Saunders RE, Derby B (2014) Inkjet printing biomaterials for tissue engineering: bioprinting. Int Mater Rev 59:430–448. https://doi.org/10.1179/1743280414Y.0000000040
21. Hospodiuk M, Dey M, Sosnoski D, Ozbolat IT (2017) The bioink: a comprehensive review on bioprintable materials. Biotechnol Adv 35:217–239. https://doi.org/10.1016/j.biotechadv.2016.12.006
22. Skoog SA, Goering PL, Narayan RJ (2014) Stereolithography in tissue engineering. J Mater Sci Mater Med 25:845–856. https://doi.org/10.1007/s10856-013-5107-y
23. Ji S, Guvendiren M (2017) Recent advances in bioink design for 3D bioprinting of tissues and organs. Front Bioeng Biotechnol 5:1–8. https://doi.org/10.3389/fbioe.2017.00023
24. Duan B, Kapetanovic E, Hockaday LA, Butcher JT (2014) Three-dimensional printed trileaflet valve conduits using biological hydrogels and human valve interstitial cells. Acta Biomater 10:1836–1846. https://doi.org/10.1016/j.actbio.2013.12.005
25. Vafaei S, Tuck C, Ashcroft I, Wildman R (2016) Surface microstructuring to modify wettability for 3D printing of nano-filled inks. Chem Eng Res Des 109:414–420. https://doi.org/10.1016/j.cherd.2016.02.004

26. Mandrycky C, Wang Z, Kim K, Kim DH (2016) 3D bioprinting for engineering complex tissues. Biotechnol Adv 34:422–434. https://doi.org/10.1016/j.biotechadv.2015.12.011
27. Xu C, Zhang M, Huang Y, Ogale A, Fu J, Markwald RR (2014) Study of droplet formation process during drop-on-demand inkjetting of living cell-laden bioink. Langmuir 30:9130–9138. https://doi.org/10.1021/la501430x
28. Rutz AL, Hyland KE, Jakus AE, Burghardt WR, Shah RN (2015) A multimaterial bioink method for 3D printing tunable, cell-compatible hydrogels. Adv Mater 27:1607–1614. https://doi.org/10.1002/adma.201405076
29. Jakus AE, Rutz AL, Shah RN (2016) Advancing the field of 3D biomaterial printing. Biomed Mater 11:014102. https://doi.org/10.1088/1748-6041/11/1/014102
30. Wüst S, Godla ME, Müller R, Hofmann S (2014) Tunable hydrogel composite with two-step processing in combination with innovative hardware upgrade for cell-based three-dimensional bioprinting. Acta Biomater 10:630–640. https://doi.org/10.1016/j.actbio.2013.10.016
31. Engler AJ, Sen S, Sweeney HL, Discher DE (2006) Matrix elasticity directs stem cell lineage specification. Cell 126:677–689. https://doi.org/10.1016/j.cell.2006.06.044
32. Ghasemi-Mobarakeh L (2015) Structural properties of scaffolds: crucial parameters towards stem cells differentiation. World J Stem Cells 7:728. https://doi.org/10.4252/wjsc.v7.i4.728
33. Velasco MA, Narváez-Tovar CA, Garzón-Alvarado DA (2015) Design, materials, and mechanobiology of biodegradable scaffolds for bone tissue engineering. Biomed Res Int 2015:1–21. https://doi.org/10.1155/2015/729076
34. Freeman FE, Kelly DJ (2017) Tuning alginate bioink stiffness and composition for controlled growth factor delivery and to spatially direct MSC Fate within bioprinted tissues. Sci Rep 7:1–12. https://doi.org/10.1038/s41598-017-17286-1
35. Jungst T, Smolan W, Schacht K, Scheibel T, Groll J (2016) Strategies and molecular design criteria for 3D printable hydrogels. Chem Rev 116:1496–1539. https://doi.org/10.1021/acs.chemrev.5b00303
36. Parsa S, Gupta M, Loizeau F, Cheung KC (2010) Effects of surfactant and gentle agitation on inkjet dispensing of living cells. Biofabrication 2:025003. https://doi.org/10.1088/1758-5082/2/2/025003
37. Gopinathan J, Noh I (2018) Recent trends in bioinks for 3D printing. Biomater Res 22:11. https://doi.org/10.1186/s40824-018-0122-1
38. Chimene D, Lennox KK, Kaunas RR, Gaharwar AK (2016) Advanced bioinks for 3D printing: a materials science perspective. Ann Biomed Eng 44:2090–2102. https://doi.org/10.1007/s10439-016-1638-y
39. Mano JF, Silva GA, Azevedo HS, Malafaya PB, Sousa RA, Silva SS, Boesel LF, Oliveira JM, Santos TC, Marques AP, Neves NM, Reis RL (2007) Natural origin biodegradable systems in tissue engineering and regenerative medicine: present status and some moving trends. J R Soc Interface 4:999–1030. https://doi.org/10.1098/rsif.2007.0220
40. Gordon TD, Schloesser L, Humphries DE, Spector M (2004) Effects of the degradation rate of collagen matrices on articular chondrocyte proliferation and biosynthesis in vitro. Tissue Eng 10(7–8):1287–1295
41. Diniz IMA, Chen C, Xu X, Ansari S, Zadeh HH, Marques MM, Shi S, Moshaverinia A (2015) Pluronic F-127 hydrogel as a promising scaffold for encapsulation of dental-derived mesenchymal stem cells. J Mater Sci Mater Med 26:1–10. https://doi.org/10.1007/s10856-015-5493-4
42. Guarino V, Caputo T, Altobelli R, Ambrosio L (2015) Degradation properties and metabolic activity of alginate and chitosan polyelectrolytes for drug delivery and tissue engineering applications. AIMS Mater Sci 2:497–502. https://doi.org/10.3934/matersci.2015.4.497
43. Kong HJ, Kaigler D, Kim K, Mooney DJ (2004) Controlling rigidity and degradation of alginate hydrogels via molecular weight distribution. Biomacromolecules 5:1720–1727. https://doi.org/10.1021/bm049879r

44. Parak A, Pradeep P, du Toit LC, Kumar P, Choonara YE, Pillay V (2018) Functionalizing bioinks for 3D bioprinting applications. Drug Discov Today 00:1–8. doi: https://doi.org/10.1016/j.drudis.2018.09.012
45. Hersel U, Dahmen C, Kessler H (2003) RGD modified polymers: biomaterials for stimulated cell adhesion and beyond. Biomaterials 24:4385–4415. https://doi.org/10.1016/S0142-9612(03)00343-0
46. Kundu J, Poole-Warren LA, Martens P, Kundu SC (2012) Silk fibroin/poly(vinyl alcohol) photocrosslinked hydrogels for delivery of macromolecular drugs. Acta Biomater 8:1720–1729. https://doi.org/10.1016/j.actbio.2012.01.004
47. De Maria C, Vozzi G, Moroni L (2017) Multimaterial, heterogeneous, and multicellular three-dimensional bioprinting. MRS Bull 42:578–584. doi: https://doi.org/10.1557/mrs.2017.165
48. Aguado BA, Mulyasasmita W, Su J, Lampe KJ, Heilshorn SC (2012) Improving viability of stem cells during syringe needle flow through the design of hydrogel cell carriers. Tissue Eng Part A 18:806–815. https://doi.org/10.1089/ten.tea.2011.0391
49. Rodriguez MJ, Brown J, Giordano J, Lin SJ, Omenetto FG, Kaplan DL (2017) Dimensional (3D) printing with in vitro and in vivo assessments. Biomaterials 117:105–115. https://doi.org/10.1016/j.biomaterials.2016.11.046.Silk
50. Cui H, Nowicki M, Fisher JP, Zhang LG (2017) 3D bioprinting for organ regeneration. Adv Healthc Mater 6:1601118. https://doi.org/10.1002/adhm.201601118
51. Norotte C, Marga FS, Niklason LE, Forgacs G (2009) Scaffold-free vascular tissue engineering using bioprinting. Biomaterials 30:5910–5917. https://doi.org/10.1016/j.biomaterials.2009.06.034
52. Seliktar D (2012) Designing cell-compatible hydrogels for biomedical applications. Science 336:1124–1128. https://doi.org/10.1126/science.1214804
53. Drury JL, Mooney DJ (2003) Hydrogels for tissue engineering: scaffold design variables and applications. Biomaterials 24:4337–4351. https://doi.org/10.1016/S0142-9612(03)00340-5
54. Zhu J, Marchant RE (2011) Design properties of hydrogel tissue-engineering scaffolds. Expert Rev Med Devices 8:607–626. https://doi.org/10.1586/erd.11.27
55. Dubbin K, Hori Y, Lewis KK, Heilshorn SC (2016) Dual-stage crosslinking of a gel-phase bioink improves cell viability and homogeneity for 3D bioprinting. Adv Healthc Mater 5:2488–2492. https://doi.org/10.1002/adhm.201600636
56. Ouyang L, Highley CB, Rodell CB, Sun W, Burdick JA (2016) 3D Printing of shear-thinning hyaluronic acid hydrogels with secondary cross-linking. ACS Biomater Sci Eng 2:1743–1751. https://doi.org/10.1021/acsbiomaterials.6b00158
57. Nichol JW, Koshy ST, Bae H, Hwang CM, Yamanlar S, Khademhosseini A (2010) Cell-laden microengineered gelatin methacrylate hydrogels. Biomaterials 31:5536–5544. https://doi.org/10.1016/j.biomaterials.2010.03.064
58. Zheng Z, Wu J, Liu M, Wang H, Li C, Rodriguez MJ, Li G, Wang X, Kaplan DL (2018) 3D bioprinting of self-standing silk-based bioink. Adv Healthc Mater 7:1701026. https://doi.org/10.1002/adhm.201701026
59. Gioffredi E, Boffito M, Calzone S, Giannitelli SM, Rainer A, Trombetta M, Mozetic P, Chiono V (2016) Pluronic F127 hydrogel characterization and biofabrication in cellularized constructs for tissue engineering applications. Proc CIRP 49:125–132. https://doi.org/10.1016/j.procir.2015.11.001
60. Guvendiren M, Molde J, Soares RMD, Kohn J (2016) Designing biomaterials for 3D printing. ACS Biomater Sci Eng 2:1679–1693. https://doi.org/10.1021/acsbiomaterials.6b00121
61. Zhang X-Z, Wu D-Q, Chu C-C (2004) Synthesis, characterization and controlled drug release of thermosensitive IPN–PNIPAAm hydrogels. Biomaterials 25:3793–3805. https://doi.org/10.1016/j.biomaterials.2003.10.065
62. Thomas D, Jessop Z, Whitaker I (2018) 3D bioprinting for reconstructive surgery: techniques and applications

63. Malda J, Visser J, Melchels FP, Jüngst T, Hennink WE, Dhert WJA, Groll J, Hutmacher DW (2013) 25th anniversary article: engineering hydrogels for biofabrication. Adv Mater 25:5011–5028. https://doi.org/10.1002/adma.201302042
64. Kreimendahl F, Köpf M, Thiebes AL, Duarte Campos DF, Blaeser A, Schmitz-Rode T, Apel C, Jockenhoevel S, Fischer H (2017) Three-dimensional printing and angiogenesis: tailored agarose-type I collagen blends comprise three-dimensional printability and angiogenesis potential for tissue-engineered substitutes. Tissue Eng Part C Methods 23:604–615. https://doi.org/10.1089/ten.tec.2017.0234
65. Forget A, Blaeser A, Miessmer F, Köpf M, Campos DFD, Voelcker NH, Blencowe A, Fischer H, Shastri VP (2017) Mechanically tunable bioink for 3D bioprinting of human cells. Adv Healthc Mater 6:1700255. https://doi.org/10.1002/adhm.201700255
66. Augst AD, Kong HJ, Mooney DJ (2006) Alginate hydrogels as biomaterials. Macromol Biosci 6:623–633. https://doi.org/10.1002/mabi.200600069
67. Chung JHY, Naficy S, Yue Z, Kapsa R, Quigley A, Moulton SE, Wallace GG (2013) Bio-ink properties and printability for extrusion printing living cells. Biomater Sci 1:763. https://doi.org/10.1039/c3bm00012e
68. Becker TA, Kipke DR (2002) Flow properties of liquid calcium alginate polymer injected through medical microcatheters for endovascular embolization. J Biomed Mater Res 61:533–540. https://doi.org/10.1002/jbm.10202
69. Kong H (2003) Designing alginate hydrogels to maintain viability of immobilized cells. Biomaterials 24:4023–4029. https://doi.org/10.1016/S0142-9612(03)00295-3
70. Kong H-J, Lee KY, Mooney DJ (2002) Decoupling the dependence of rheological/mechanical properties of hydrogels from solids concentration. Polymer (Guildf) 43:6239–6246. https://doi.org/10.1016/S0032-3861(02)00559-1
71. Kuo CK, Ma PX (2001) Ionically crosslinked alginate hydrogels as scaffolds for tissue engineering: Part 1. Structure, gelation rate and mechanical properties. Biomaterials 22:511–521. https://doi.org/10.1016/S0142-9612(00)00201-5
72. Jia J, Richards DJ, Pollard S, Tan Y, Rodriguez J, Visconti RP, Trusk TC, Yost MJ, Yao H, Markwald RR, Mei Y (2014) Engineering alginate as bioink for bioprinting. Acta Biomater 10:4323–4331. https://doi.org/10.1016/j.actbio.2014.06.034
73. Ferreira AM, Gentile P, Chiono V, Ciardelli G (2012) Collagen for bone tissue regeneration. Acta Biomater 8:3191–3200. https://doi.org/10.1016/j.actbio.2012.06.014
74. Billiet T, Gevaert E, De Schryver T, Cornelissen M, Dubruel P (2014) The 3D printing of gelatin methacrylamide cell-laden tissue-engineered constructs with high cell viability. Biomaterials 35:49–62. doi: https://doi.org/10.1016/j.biomaterials.2013.09.078
75. Gruene M, Pflaum M, Hess C, Diamantouros S, Schlie S, Deiwick A, Koch L, Wilhelmi M, Jockenhoevel S, Haverich A, Chichkov B (2011) Laser printing of three-dimensional multicellular arrays for studies of cell–cell and cell–environment interactions. Tissue Eng Part C Methods 17:973–982. https://doi.org/10.1089/ten.tec.2011.0185
76. Highley CB, Prestwich GD, Burdick JA (2016) Recent advances in hyaluronic acid hydrogels for biomedical applications. Curr Opin Biotechnol 40:35–40. https://doi.org/10.1016/j.copbio.2016.02.008
77. Poldervaart MT, Goversen B, de Ruijter M, Abbadessa A, Melchels FPW, Öner FC, Dhert WJA, Vermonden T, Alblas J (2017) 3D bioprinting of methacrylated hyaluronic acid (MeHA) hydrogel with intrinsic osteogenicity. PLoS One 12:1–15. doi: https://doi.org/10.1371/journal.pone.0177628
78. Skardal A, Zhang J, Prestwich GD (2010) Bioprinting vessel-like constructs using hyaluronan hydrogels crosslinked with tetrahedral polyethylene glycol tetracrylates. Biomaterials 31:6173–6181. https://doi.org/10.1016/j.biomaterials.2010.04.045
79. Nguyen D, Hägg DA, Forsman A, Ekholm J, Nimkingratana P, Brantsing C, Kalogeropoulos T, Zaunz S, Concaro S, Brittberg M, Lindahl A, Gatenholm P, Enejder A, Simonsson S (2017) Cartilage tissue engineering by the 3D bioprinting of ips cells in a nanocellulose/alginate bioink. Sci Rep 7:658. https://doi.org/10.1038/s41598-017-00690-y

80. Markstedt K, Mantas A, Tournier I, Martínez Ávila H, Hägg D, Gatenholm P (2015) 3D bioprinting human chondrocytes with nanocellulose–alginate bioink for cartilage tissue engineering applications. Biomacromolecules 16:1489–1496. https://doi.org/10.1021/acs.biomac.5b00188
81. Lott JR, McAllister JW, Arvidson SA, Bates FS, Lodge TP (2013) Fibrillar structure of methylcellulose hydrogels. Biomacromolecules 14:2484–2488. https://doi.org/10.1021/bm400694r
82. Kobayashi K, Huang C, Lodge TP (1999) Thermoreversible gelation of aqueous methylcellulose solutions. Macromolecules 32:7070–7077. https://doi.org/10.1021/ma990242n
83. Thirumala S, Gimble J, Devireddy R (2013) Methylcellulose based thermally reversible hydrogel system for tissue engineering applications. Cell 2:460–475. https://doi.org/10.3390/cells2030460
84. Kundu B, Rajkhowa R, Kundu SC, Wang X (2013) Silk fibroin biomaterials for tissue regenerations. Adv Drug Deliv Rev 65:457–470. https://doi.org/10.1016/j.addr.2012.09.043
85. Floren M, Bonani W, Dharmarajan A, Motta A, Migliaresi C, Tan W (2016) Human mesenchymal stem cells cultured on silk hydrogels with variable stiffness and growth factor differentiate into mature smooth muscle cell phenotype. Acta Biomater 31:156–166. https://doi.org/10.1016/j.actbio.2015.11.051
86. Rodriguez MJ, Brown J, Giordano J, Lin SJ, Omenetto FG, Kaplan DL (2017) Silk based bioinks for soft tissue reconstruction using 3-dimensional (3D) printing with in vitro and in vivo assessments. Biomaterials 117:105–115. https://doi.org/10.1016/j.biomaterials.2016.11.046
87. Wang X, Partlow B, Liu J, Zheng Z, Su B, Wang Y, Kaplan DL (2015) Injectable silk-polyethylene glycol hydrogels. Acta Biomater 12:51–61. https://doi.org/10.1016/j.actbio.2014.10.027
88. Frantz C, Stewart KM, Weaver VM (2010) The extracellular matrix at a glance. J Cell Sci 123:4195–4200. https://doi.org/10.1242/jcs.023820
89. Jung JP, Bhuiyan DB, Ogle BM (2016) Solid organ fabrication: comparison of decellularization to 3D bioprinting. Biomater Res 20:27. https://doi.org/10.1186/s40824-016-0074-2
90. Jang J, Park HJ, Kim SW, Kim H, Park JY, Na SJ, Kim HJ, Park MN, Choi SH, Park SH, Kim SW, Kwon SM, Kim PJ, Cho DW (2017) 3D printed complex tissue construct using stem cell-laden decellularized extracellular matrix bioinks for cardiac repair. Biomaterials 112:264–274. https://doi.org/10.1016/j.biomaterials.2016.10.026
91. Lutolf MP, Hubbell JA (2005) Synthetic biomaterials as instructive extracellular microenvironments for morphogenesis in tissue engineering. Nat Biotechnol 23:47–55. https://doi.org/10.1038/nbt1055
92. Kang HW, Lee SJ, Ko IK, Kengla C, Yoo JJ, Atala A (2016) A 3D bioprinting system to produce human-scale tissue constructs with structural integrity. Nat Biotechnol 34:312–319. https://doi.org/10.1038/nbt.3413
93. Khattak SF, Bhatia SR, Roberts SC (2005) Pluronic F127 as a cell encapsulation material: utilization of membrane-stabilizing agents. Tissue Eng 11:974–983. https://doi.org/10.1089/ten.2005.11.974
94. Müller M, Becher J, Schnabelrauch M, Zenobi-Wong M (2015) Nanostructured pluronic hydrogels as bioinks for 3D bioprinting. Biofabrication 7(3):035006. https://doi.org/10.1088/1758-5090/7/3/035006
95. Hockaday LA, Kang KH, Colangelo NW, Cheung PYC, Duan B, Malone E, Wu J, Girardi LN, Bonassar LJ, Lipson H, Chu CC, Butcher JT (2012) Rapid 3D printing of anatomically accurate and mechanically heterogeneous aortic valve hydrogel scaffolds. Biofabrication 4:035005. https://doi.org/10.1088/1758-5082/4/3/035005
96. Gao G, Yonezawa T, Hubbell K, Dai G, Cui X (2015) Inkjet-bioprinted acrylated peptides and PEG hydrogel with human mesenchymal stem cells promote robust bone and cartilage formation with minimal printhead clogging. Biotechnol J 10:1568–1577. https://doi.org/10.1002/biot.201400635
97. Kesti M, Müller M, Becher J, Schnabelrauch M, D'Este M, Eglin D, Zenobi-Wong M (2015) A versatile bioink for three-dimensional printing of cellular scaffolds based on thermally

and photo-triggered tandem gelation. Acta Biomater 11:162–172. https://doi.org/10.1016/j.actbio.2014.09.033
98. Wu D, Yu Y, Tan J, Huang L, Luo B, Lu L, Zhou C (2018) 3D bioprinting of gellan gum and poly (ethylene glycol) diacrylate based hydrogels to produce human-scale constructs with high-fidelity. Mater Des 160:486–495. https://doi.org/10.1016/j.matdes.2018.09.040
99. Wilson SA, Cross LM, Peak CW, Gaharwar AK (2017) Shear-thinning and thermo-reversible nanoengineered inks for 3D bioprinting. ACS Appl Mater Interfaces 9:43449–43458. https://doi.org/10.1021/acsami.7b13602
100. Bertassoni LE, Cardoso JC, Manoharan V, Cristino AL, Bhise NS, Araujo WA, Zorlutuna P, Vrana NE, Ghaemmaghami AM, Dokmeci MR, Khademhosseini A (2014) Direct-write bioprinting of cell-laden methacrylated gelatin hydrogels. Biofabrication 6:024105. https://doi.org/10.1088/1758-5082/6/2/024105
101. Shim JH, Kim JY, Park M, Park J, Cho DW (2011) Development of a hybrid scaffold with synthetic biomaterials and hydrogel using solid freeform fabrication technology. Biofabrication 3:034102. https://doi.org/10.1088/1758-5082/3/3/034102
102. Hinton TJ, Jallerat Q, Palchesko RN, Park JH, Grodzicki MS, Shue H-J, Ramadan MH, Hudson AR, Feinberg AW (2015) Three-dimensional printing of complex biological structures by freeform reversible embedding of suspended hydrogels. Sci Adv 1:e1500758. https://doi.org/10.1126/sciadv.1500758
103. Bhattacharjee T, Zehnder SM, Rowe KG, Jain S, Nixon RM, Sawyer WG, Angelini TE (2015) Writing in the granular gel medium. Sci Adv 1:e1500655. https://doi.org/10.1126/sciadv.1500655
104. Jenkins AD, Kratochvíl P, Stepto RFT, Suter UW (1996) Glossary of basic terms in polymer science (IUPAC Recommendations 1996). Pure Appl Chem 68:2287–2311. https://doi.org/10.1351/pac199668122287
105. Lin P, Ma S, Wang X, Zhou F (2015) Molecularly engineered dual-crosslinked hydrogel with ultrahigh mechanical strength, toughness, and good self-recovery. Adv Mater 27:2054–2059. https://doi.org/10.1002/adma.201405022
106. Chen H, Liu Y, Ren B, Zhang Y, Ma J, Xu L, Chen Q, Zheng J (2017) Super bulk and interfacial toughness of physically crosslinked double-network hydrogels. Adv Funct Mater 27:1–10. https://doi.org/10.1002/adfm.201703086
107. Pulieri E, Chiono V, Ciardelli G, Vozzi G, Ahluwalia A, Domenici C, Vozzi F, Giusti P (2008) Chitosan/gelatin blends for biomedical applications. J Biomed Mater Res A 86A:311–322. https://doi.org/10.1002/jbm.a.31492
108. Naseri N, Deepa B, Mathew AP, Oksman K, Girandon L (2016) Nanocellulose-based interpenetrating polymer network (IPN) hydrogels for cartilage applications. Biomacromolecules 17:3714–3723. https://doi.org/10.1021/acs.biomac.6b01243
109. Hoare TR, Kohane DS (2008) Hydrogels in drug delivery: progress and challenges. Polymer (Guildf) 49:1993–2007. https://doi.org/10.1016/j.polymer.2008.01.027
110. Reis AV, Guilherme MR, Moia TA, Mattoso LHC, Muniz EC, Tambourgi EB (2008) Synthesis and characterization of a starch-modified hydrogel as potential carrier for drug delivery system. J Polym Sci A Polym Chem 46:2567–2574. https://doi.org/10.1002/pola.22588
111. Muzzarelli RAA (2009) Genipin-crosslinked chitosan hydrogels as biomedical and pharmaceutical aids. Carbohydr Polym 77:1–9. https://doi.org/10.1016/j.carbpol.2009.01.016
112. Bajaj P, Schweller RM, Khademhosseini A, West JL, Bashir R (2014) 3D biofabrication strategies for tissue engineering and regenerative medicine. Annu Rev Biomed Eng 16:247–276. https://doi.org/10.1146/annurev-bioeng-071813-105155
113. Roland CM (2015) Interpenetrating polymer networks (IPN): structure and mechanical behavior. In: Encyclopedia of polymeric nanomaterials. Springer, Berlin, pp 1004–1011
114. Pescosolido L, Schuurman W, Malda J, Matricardi P, Alhaique F, Coviello T, van Weeren PR, Dhert WJA, Hennink WE, Vermonden T (2011) Hyaluronic acid and dextran-based semi-IPN hydrogels as biomaterials for bioprinting. Biomacromolecules 12:1831–1838. doi: https://doi.org/10.1021/bm200178w

115. Chatterjee U, Jewrajka SK, Guha S (2009) Dispersion of functionalized silver nanoparticles in polymer matrices: stability, characterization, and physical properties. Polym Compos 30:827–834. https://doi.org/10.1002/pc.20655
116. Liu HL, Dai SA, Fu KY, Hsu SH (2010) Antibacterial properties of silver nanoparticles in three different sizes and their nanocomposites with a new waterborne polyurethane. Int J Nanomed 5:1017–1028. https://doi.org/10.2147/IJN.S14572
117. Deka H, Karak N, Kalita RD, Buragohain AK (2010) Bio-based thermostable, biodegradable and biocompatible hyperbranched polyurethane/Ag nanocomposites with antimicrobial activity. Polym Degrad Stab 95:1509–1517. https://doi.org/10.1016/j.polymdegradstab.2010.06.017
118. Zhu K, Shin SR, van Kempen T, Li Y-C, Ponraj V, Nasajpour A, Mandla S, Hu N, Liu X, Leijten J, Lin Y-D, Hussain MA, Zhang YS, Tamayol A, Khademhosseini A (2017) Gold nanocomposite bioink for printing 3D cardiac constructs. Adv Funct Mater 27:1605352. doi: https://doi.org/10.1002/adfm.201605352
119. Cobley CM, Chen J, Cho EC, Wang LV, Xia Y (2011) Gold nanostructures: a class of multifunctional materials for biomedical applications. Chem Soc Rev 40:44–56. https://doi.org/10.1039/b821763g
120. Pekkanen AM, Mondschein RJ, Williams CB, Long TE (2017) 3D printing polymers with supramolecular functionality for biological applications. Biomacromolecules 18:2669–2687. https://doi.org/10.1021/acs.biomac.7b00671
121. Yang L, Tan X, Wang Z, Zhang X (2015) Supramolecular polymers: historical development, preparation, characterization, and functions. Chem Rev 115:7196–7239. https://doi.org/10.1021/cr500633b
122. Clarke DE, Pashuck ET, Bertazzo S, Weaver JVM, Stevens MM (2017) Self-healing, self-assembled β-sheet peptide–poly(γ-glutamic acid) hybrid hydrogels. J Am Chem Soc 139:7250–7255. https://doi.org/10.1021/jacs.7b00528
123. Bahlmann LC, Fokina A, Shoichet MS (2017) Dynamic bioengineered hydrogels as scaffolds for advanced stem cell and organoid culture. MRS Commun 7:472–486. https://doi.org/10.1557/mrc.2017.72
124. Radu-Wu LC, Yang J, Wu K, Kopeček J (2009) Self-assembled hydrogels from poly[*N*-(2-hydroxypropyl)methacrylamide] grafted with β-sheet peptides. Biomacromolecules 10:2319–2327. https://doi.org/10.1021/bm9005084
125. Elder AN, Dangelo NM, Kim SC, Washburn NR (2011) Conjugation of β-sheet peptides to modify the rheological properties of hyaluronic acid. Biomacromolecules 12:2610–2616. https://doi.org/10.1021/bm200393k
126. Highley CB, Rodell CB, Burdick JA (2015) Direct 3D printing of shear-thinning hydrogels into self-healing hydrogels. Adv Mater 27:5075–5079. https://doi.org/10.1002/adma.201501234
127. Li C, Faulkner-Jones A, Dun AR, Jin J, Chen P, Xing Y, Yang Z, Li Z, Shu W, Liu D, Duncan RR (2015) Rapid formation of a supramolecular polypeptide-DNA hydrogel for in situ three-dimensional multilayer bioprinting. Angew Chem Int Ed Engl 54:3957–3961. https://doi.org/10.1002/anie.201411383
128. Dubbin K, Tabet A, Heilshorn SC (2017) Quantitative criteria to benchmark new and existing bio-inks for cell compatibility. Biofabrication 9:044102. https://doi.org/10.1088/1758-5090/aa869f
129. Wang H, Heilshorn SC (2015) Adaptable hydrogel networks with reversible linkages for tissue engineering. Adv Mater 27:3717–3736. https://doi.org/10.1002/adma.201501558
130. Pashuck ET (2018) Designing self-assembling biomaterials with controlled mechanical and biological performance. In: Self-assembling biomaterials, pp 7–26. https://doi.org/10.1016/B978-0-08-102015-9.00002-2
131. Jiang Y, Chen J, Deng C, Suuronen EJ, Zhong Z (2014) Click hydrogels, microgels and nanogels: emerging platforms for drug delivery and tissue engineering. Biomaterials 35:4969–4985. https://doi.org/10.1016/j.biomaterials.2014.03.001

132. Jiang W, Li M, Chen Z, Leong KW (2016) Cell-laden microfluidic microgels for tissue regeneration. Lab Chip 16:4482–4506. https://doi.org/10.1039/C6LC01193D
133. Griffin DR, Weaver WM, Scumpia PO, Di Carlo D, Segura T (2015) Accelerated wound healing by injectable microporous gel scaffolds assembled from annealed building blocks. Nat Mater 14:737–744. doi: https://doi.org/10.1038/nmat4294
134. Mealy JE, Chung JJ, Jeong HH, Issadore D, Lee D, Atluri P, Burdick JA (2018) Injectable granular hydrogels with multifunctional properties for biomedical applications. Adv Mater 30:1–7. https://doi.org/10.1002/adma.201705912

Suggested Reading

Chimene D, Lennox KK, Kaunas RR, Gaharwar AK (2016) Advanced bioinks for 3D printing: a materials science perspective. Ann Biomed Eng 44(6):2090–2102
Cui H, Nowicki M, Fisher JP, Zhang LG (2017) 3D bioprinting for organ regeneration. Adv Healthc Mater 6(1). https://doi.org/10.1002/adhm.201601118
Gopinathan J, Noh I (2018) Recent trends in bioinks for 3D printing. Biomater Res 22(1):1–15
Ji S, Guvendiren M (2017) Recent advances in bioink design for 3D bioprinting of tissues and organs. Front Bioeng Biotechnol 5:1–8
Hospodiuk M, Dey M, Sosnoski D, Ozbolat IT (2017) The bioink: a comprehensive review on bioprintable materials. Biotechnol Adv 35(2):217–239
Jungst T, Smolan W, Schacht K, Scheibel T, Groll J (2016) Strategies and molecular design criteria for 3D printable hydrogels. Chem Rev 116(3):1496–1539

Chapter 3
Potential Clinical Applications of Three-Dimensional Bioprinting

Ippokratis Pountos, Nazzar Tellisi, and Nureddin Ashammakhi

Abstract Three-dimensional (3D) bioprinting aims to construct complex personalized living tissues mimicking the native tissues. This chapter presents the current advances in 3D bioprinting. Available evidence revealed promising results in potential applications for the regeneration of musculoskeletal, cardiovascular, dermal, and neural tissues. These applications comprise a developing field. However, there are still barriers that hamper further expansion of this technology. Such challenges involve the reliable mechanical properties, size limitations, integration of transplanted grafts, and safeguarding of safety throughout the process of 3D printing and resulting constructs.

Keywords 3D bioprinting · Bone · Mesenchymal stem cells · Cartilage · Blood vessels · Cardiovascular tissue

---

I. Pountos (✉) · N. Tellisi
Academic Department of Trauma and Orthopaedics, Leeds Teaching Hospitals, University of Leeds, Leeds, UK

Chapel Allerton Hospital, Leeds Teaching Hospitals, Leeds, UK
e-mail: pountos@doctors.org.uk

N. Ashammakhi
Center for Minimally Invasive Therapeutics (C-MIT), University of California Los Angeles, Los Angeles, CA, USA

California NanoSystems Institute (CNSI), University of California Los Angeles, Los Angeles, CA, USA

Department of Radiological Sciences, David Geffen School of Medicine, University of California Los Angeles, Los Angeles, CA, USA

Biotechnology Research Center, Libyan Authority for Research, Science and Technology, Tripoli, Libya
e-mail: n.ashammakhi@ucla.edu

M. Guvendiren (ed.), *3D Bioprinting in Medicine*,
https://doi.org/10.1007/978-3-030-23906-0_3

3.1 Introduction

Bioprinting is the process of combining cellular and non-cellular components in bioinks to produce three-dimensional (3D) constructs that can mimic or be used to reconstruct human tissues. This technology is based on 'additive fabrication' of layers to achieve 3D fabrication of tissues that can replicate the hierarchical structure and cell composition of native tissues. Three-dimensional bioprinting as a concept is far superior to currently available tissue engineering approaches that involve the loading of cells and/or growth factors into scaffolds. This technology offers the ability to fabricate 3D tissue structures with high precision, fidelity, and stability at the human clinical scale [1, 2]. The creation of complex tissue architectures with heterogeneous compositions has the potential to revolutionize the transplantation of tissues. Since the medical community realized its potential, 3D bioprinting has captured significant interest, especially over the last decade [3, 4].

In brief, 3D bioprinting uses three common printing technologies, microextrusion, inkjet, and laser-assisted bioprinting methods (Fig. 3.1) [5]. It also involves three distinct steps: (1) pre-processing, (2) processing, and (3) post-processing. Pre-processing involves the creation of a computer-aided design of the tissue. Magnetic resonance imaging (MRI) or computed tomography (CT) scans can be utilized for computer-controlled 3D printing using appropriate bioinks [6]. Cells can be harvested and used fresh or can be manipulated ex vivo. Depending on the tissue of interest, our armamentarium includes a variety of bioinks and hardware. This is often a crucial element that can influence the quality and survival of the graft. Processing is the actual bioprinting of the tissue, while post-processing involves the brief incubation of the tissue or graft in a bioreactor. To minimize the ex vivo manip-

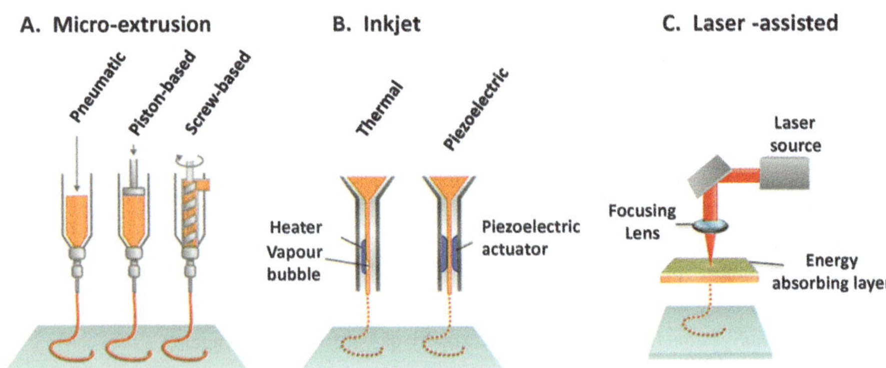

Fig. 3.1 Three-dimensional bioprinting techniques. The three most common 3D bioprinting techniques: (**a**) microextrusion, (**b**) inkjet, and (**c**) laser-assisted 3D bioprinting. The microextrusion technique can be (1) pneumatic, (2) piston-based mechanical, or (3) screw-based. Inkjet technique can be either thermal or piezoelectric. (**a**, **b**) are adapted from Malda et al. with permission from John Wiley and Sons [5]. (**c**) is adapted from Keriquel et al. [28, 29] with permission from Nature Publishing Group [29]

ulation of these constructs, in situ bioprinting can be performed, i.e. 3D bioprinting directly on the defect site (Fig. 3.2) [7] Using these approaches, a number of tissues have been created with promising results. Tissues of musculoskeletal origin, neural or vascular structures, skin, and other have been developed over the years [8, 9].

In this chapter, we aim to present some of the current and numerous potential clinical applications of 3D bioprinting. Challenges and future developments of 3D bioprinting are also discussed here in this chapter.

3.2 Bone

Bone is a unique tissue that provides stability to the whole body and performs other functions like haematopoiesis, locomotion, and homeostasis of important elements of the human body. Bone can be fractured following trauma, and bony defects can occur following severe injuries, tumours, or other pathologies [10–12]. In addition, some fractures and bony injuries fail to heal. It is estimated that 5–10% of long bone fractures will end up in nonunion [12, 13]. Tissue engineering approaches have focused on assisting bone regeneration as an attempt to either upregulate the overall healing process in high-risk cases or provide the required scaffold, cells, and osteo-inductive factors in cases of bone loss [10–14]. With the advances made in 3D bio-printing, several attempts to create bone were made. The main challenges remain the selection of materials with optimal rheological properties, biocompatibility,

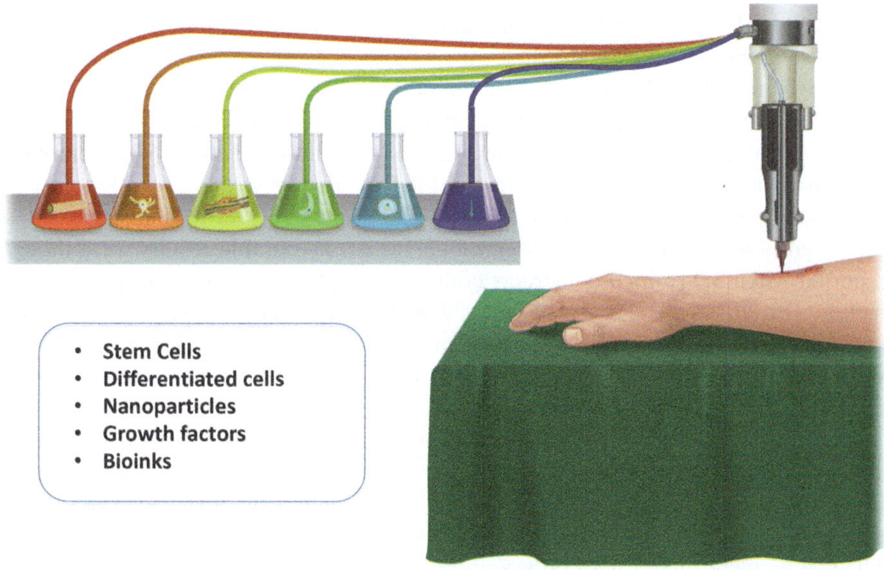

Fig. 3.2 In situ 3D bioprinting of skin. On-demand personalized bioprinting directly on the defect with bioinks matching the reconstructed tissues

osteoconductivity, and capacity of the graft to be incorporated and remodelled to normal bone [1].

In vitro studies analysed the effectiveness of a number of bioinks for the fabrication of bone (Table 3.1) [1, 6, 15–34]. Some investigators used materials such as poly-ε-caprolactone (PCL) and poly(lactide-*co*-glycolide) (PLGA). Although these are mechanically stable, they have limited osteoconductive properties [19, 20]. A number of alternative materials were explored such as tricalcium phosphates, hydroxyapatite, and bioactive glass [15, 21–23]. In the works of Poldervaart et al., methacrylated hyaluronic acid (HA) with human mesenchymal stem cells (MSCs) was used [15]. The cellular viability was 64% after 21 days of culture, and osteogenic differentiation of MSCs occurred spontaneously in hydrogels. The osteogenic differentiation increased with the addition of bone morphogenetic protein-2 (BMP-2) to the culture medium [15]. In a similar study using gelatin and alginate bioinks with human adipose-derived stem cells, high cellular viability levels were noted with the expression of important osteogenic markers [16]. Similar composite materials were used in a number of studies with favourable results. The combination, for example, of thermo-responsive hydrogels with collagen type I improved the mechanical properties of the constructs [18]. Other investigators used bioactive glass, microcarriers, polymers, and polyethylene, and they obtained similar results when attempting to increase the overall mechanical stability of the 3D printed hydrogels [24–26]. Comparable results were described with other base bioinks, for instance, adding microcarriers to gelatin methacryloyl (GelMA) [19]. A different approach is to load osteogenic growth factors into the bioink. In a study on BMP-2 loaded gelatin, release kinetics and bioactivity showed continuous release of BMP-2 for 3 weeks after bioprinting [17]. Using the aforementioned technologies it was possible to fabricate a whole human mandible as well as calvarial bone, cartilage, and skeletal muscle [Fig. 3.3] [6].

In vivo animal studies have also shown promising results [16, 17]. In a segmental tibial defect model, the application of nano-hydroxyapatite (nHAp)/PCL resulted in the formation of dense bone tissue around the scaffold at 8 weeks postoperatively [22]. In a similar model, of rabbit femoral defects that were treated with poly(D,L-lactide-*co*-glycolide) and β-tricalcium phosphate (β-TCP) nanocomposites, increased bone formation was observed [27]. Less favourable results were though reported when 3D printed constructs composed of TCP and poly(L-lactide-*co*-D,L-lactide) (PLDLLA)-TCP-PCL scaffolds were implanted in ovine segmental defects [1]. At 12 weeks postoperatively, only minor external callus and bone formation were observed, suggesting that adding a biologically active stimulus such as a BMP might be required [1].

In situ 3D bioprinting of bone has been also proposed by a limited number of studies. In the works of Keriquel et al., the feasibility of a laser bioprinter adapted for in vivo use on calvarial defects in mice was studied [Fig. 3.4] [28]. The investigators used nHAp to fill the defects and they followed up the animals for 3 months. The results were mixed with only a proportion of the defects was filled with bone tissue. The same group used this technology employing mesenchymal stromal cells in different arrangements and geometries within a scaffold composed of nHA and

Table 3.1 Selected studies showing evidence of successful bone 3D bioprinting

Author, Year	Printer	Design	Scaffold/Bioink	Cells	Outcome
Keriquel et al., 2010 [28]	Laser (In situ)	In vivo	Nano hydroxyapatite (nHA)	No cells	• In vivo bioprinting is possible. • No effect to animal's brain. Bone formation only occurred in some defects.
Kim et al., 2012 [22]	Extrusion	In vivo	DL-PLGA and β-tricalcium phosphate (β-TCP) nanocomposites	No cells	• Scaffolds integrated with the host bone and were biocompatible.
Poldervaart et al., 2013 [17]	Extrusion	In vitro and in vivo	Gelatin loaded BMP-2 and alginate	Goat multipotent stromal cells	• Controlled release of BMP-2 from the scaffold was noted.
Du et al., 2015 [34]	Extrusion	In vitro	Methacrylamide gelatin scaffold with collagen microfibers and BMP-2	MSCs	• BMP-2 was able to be controllably released. • MSCs showed high cell viability (>90%) during printing. • CBD-BMP2-collagen microfibers induced BMSC differentiation into osteocytes within 14 days in culture.
Duarte Campos et al., 2016 [18]	Extrusion	In vitro	Collagen type I in polysaccharide-based hydrogels	Human MSCs	• MSC not only survive the 3D-bioprinting process but also maintain the mesenchymal phenotype.
Wang et al., 2016 [16]	Extrusion	In vitro and in vivo	Gelatin and alginate	Human adipose-derived stem cells	• Cell viability of 89% on day 1 after printing. • The expression levels of RUNX2, OSX, and OCN were significantly increased on days 7 and 14 after printing. • Bone matrix formation in the 3D bioprinted constructs noted in vivo.

(continued)

Table 3.1 (continued)

Author, Year	Printer	Design	Scaffold/Bioink	Cells	Outcome
Bendtsen et al., 2017 [30]	Extrusion	In vitro	Alginate-polyvinyl alcohol (PVA)-hydroxyapatite (HA) hydrogel	Mouse calvaria cells (MC3T3)	• Construct remained stable for 2 weeks and high cellular viability was noted.
Demirtas et al., 2017 [31]	Extrusion	In vitro	Chitosan solution with nanostructured bone-like hydroxyapatite	MC3T3-E1 pre-osteoblast	• Stable construct that preserved cell viability and allowed osteogenic differentiation in culture.
Keriquel et al., 2017 [29]	Laser (in situ)	In vivo (critical size defect)	Nano hydroxyapatite (nHA) and collagen	MSCs	• This technology can print complex structures and favour bone regeneration. • Cell geometries and cell arrangements have a significant impact on bone regeneration.
Neufurth et al., 2017 [32]	Extrusion	In vitro	Amorphous microparticles prepared from Ca^{2+} and the physiological inorganic polymer, polyphosphate fortified by mixing with poly-ε-caprolactone	Human bone-related SaOS-2	• Scaffold was capable of attracting and promoting the growth of human bone SaOS-2 cells.
Poldervaart et al., 2017 [15]	Extrusion	In vitro	Methacrylated hyaluronic acid (MeHA) gel	Human MSCs	• Osteogenic differentiation of MSCs occurred spontaneously and further enhanced with the addition of BMP-2. Cell viability remained 64.4% after 21 days of culture.
Zhang et al., 2017 [33]	Extrusion	In vitro	β-tricalcium phosphate bioceramic scaffolds containing silver nanoparticles on graphene oxide	Rabbit bone marrow stromal cells	• Excellent antibacterial activity accelerated osteogenic differentiation of the cells.

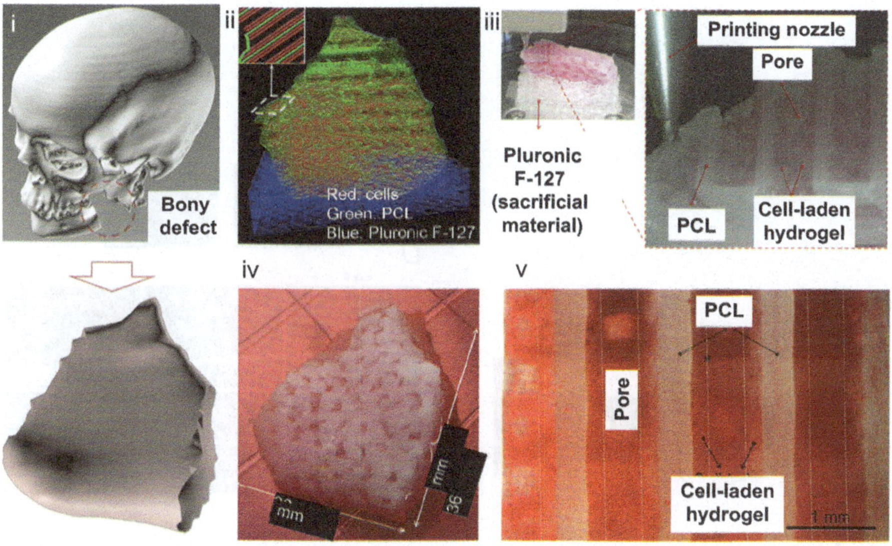

Fig. 3.3 Reconstruction of a human mandible graft. (**i**) Using CT images a 3D CAD model was created. (**ii**) Reconstruction of the bone defect 3D architecture: Green, blue, and red lines represent the paths used to dispense various inks (PCL, Pluronic F-127, and cell-laden hydrogel, respectively). (**iii**) The patterning of a construct layer using 3D bioprinting. (**iv**) Appearance of the construct in culture after 28 days in osteogenic medium. It was cultured in osteogenic medium for 28 days. (**v**) Calcium deposition following osteogenic differentiation in the printed construct was evident by Alizarin Red S staining. Figure 3.3 was reproduced from Kang et al. [87] with permission from Nature Publishing Group

collagen [29]. The investigators showed that this technology can produce favourable results in the setting of large bone defects. It also demonstrated that a disc configuration with MSCs had the best results in bone healing and regeneration [29].

3.3 Cartilage

Cartilage damage and osteoarthritis affect millions of people worldwide. In the USA alone, osteoarthritis affects 37% of the population over 65 causing significant morbidity and reduction in quality of life [35]. Our current approach to the management of osteoarthritis is carrying out joint arthroplasty or fusion, while cartilage regeneration techniques are still in their infancy and with controversial results [36]. However, cartilage is a unique tissue and possibly ideal target for 3D bioprinting applications, as it does not require blood vessels. Many studies currently show that 3D bioprinted cartilage could be a solution to the cartilage loss and may offer better treatment for arthritis [Table 3.2] [37–54].

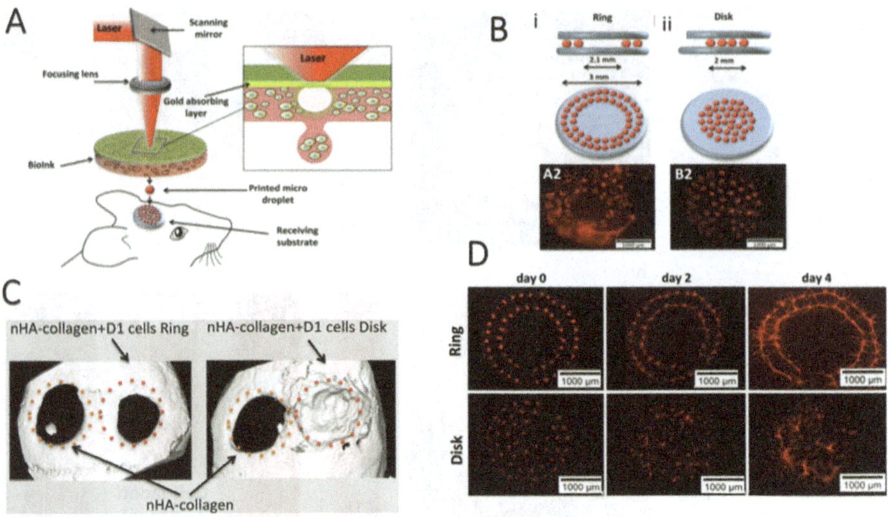

Fig. 3.4 Laser 3D bioprinting on osteoconductive discs. Osteoprogenitor cells were printed on nHAp collagen discs and subsequently used in the treatment of experimental calvarial defects in mice (**a**). Cells were printed at the peripheral (i) or central (ii) areas of the discs (**b**). Immediately after printing fluorescence images of peripherally (A2) and centrally (B2) printed tomato-positive cells, (**c**) Microtomography (μCT) images, 2 months after surgery, showing increased osteogenic activity for the defects where cells were applied at the central area rather than the periphery. No bone formation was noted in defects where no cells were applied. (**d**) Day 0, 2, and 4 fluorescence images of centrally and peripherally printed tomato-positive D1 cells. Figure is reproduced from Keriquel et al. [28, 29] with permission from Nature Publishing Group

In vitro studies have shown that 3D bioprinted cartilage is feasible [38–40]. Three-dimensional bioprinted cartilage can have long-term stability and mechanical integrity [39, 40]. In addition, the cells used in the bioprinting process have shown acceptable viability levels and remain functional after the fabrication of cartilage [38–40]. Daly et al. compared a number of bioinks for their capacity to support chondrocytes and for ultimately developing hyaline cartilage [42]. They suggest that alginate and agarose hydrogels were superior in developing hyaline-like cartilage as compared to GelMA and BioINK™. The latter resulted in the formation of tissue with cellular phenotype resembling fibrocartilage even though all bioinks supported the cells and achieved high viability. In another model of knee osteoarthritis, silk fibroin with gelatin combined with BMSC-specific-affinity peptide showed promising results [37]. The authors suggested that silk fibroin and gelatin can greatly balance the mechanical properties and degradation rate to match the newly formed cartilage [37]. A different approach involving on-demand personalized biofabrication of grafts was also explored. Di Bella et al. investigated the effectiveness of a hand-held bioprinter on critical size osteochondral defects in sheeps [41]. The bioink comprised gelatin methyacrylamide, HA methacrylate hydrogel, and MSCs [41]. The investigators reported better macroscopic and microscopic appearance of the resulting tissue was comparable to that achieved with microfrac-

Table 3.2 Selected studies showing the current evidence on 3D bioprinting of cartilage

Author, Year	Printer	Design	Scaffold/Bioink	Cells	Outcome
Cui et al., 2012 [38]	Inject	In vitro	Poly(ethylene glycol) dimethacrylate (PEGDMA)	Human chondrocytes	• High cellular viability with preservation of function. • Promising anatomic cartilage engineering using 3D bioprinting technology.
Schuurman et al., 2013 [49]	Extrusion	In vitro	Gelatin-methacrylamide with/without ε-polycaprolactone or HA	Human chondrocytes	• When gelatin-methacrylamide is combined with HA and/or a reinforcing support structure, such as PCL, gelMA can be fabricated into layered hydrogel structures, which could aid in the engineering of human cartilage.
Xu et al., 2013 [40]	Hybrid (inkjet and electrospinning system)	In vitro and in vivo	Polycaprolactone fibers and chondrocytes suspended in a fibrin-collagen hydrogel	Rabbit chondrocytes	• 80% viability 1 week after printing was noted. • Cells proliferated and maintained their basic biological properties. • Constructs formed cartilage-like tissues both in vitro and in vivo as evidenced by the deposition of type II collagen and glycosaminoglycans. .
Kundu et al., 2015 [39]	Extrusion	In vitro and in vivo	Polycaprolactone and chondrocyte cell-encapsulated alginate hydrogel	Human chondrocytes	• Enhanced cartilage tissue and type II collagen fibril formation at 4 weeks following implantation in vivo were observed.
Markstedt et al., 2015 [50]	Extrusion	In vitro	Nanofibrillated cellulose	Human chondrocytes	• Cell viability of 73% and 86% after 1 and 7 days, respectively, was noted.

(continued)

Table 3.2 (continued)

Author, Year	Printer	Design	Scaffold/Bioink	Cells	Outcome
Costantini et al., 2016 [51]	Extrusion	In vitro	Gelatin methacrylamide, chondroitin sulphate amino ethyl methacrylate, and HA methacrylate.	Human MSCs	• Enhanced viability and chondrogenic differentiation of BM-MSCs was noted.
Ren et al., 2016 [52]	Extrusion	In vitro	Collagen type II hydrogel	Rabbit chondrocytes	• ECM production was positively correlated with the total cell density.
Daly et al., 2016 [42]	Extrusion	In vitro	Agarose, alginate, GelMA, and BioINK™	MSCs	• High viability levels with all bioinks were reported. • GelMA and BioINK™ resulted in developing a more fibrocartilage-like tissue.
Nguyen et al., 2017 [53]	Extrusion	In vitro	Nanofibrillated cellulose composite with alginate or HA	Human chondrocytes	• Cell viability, pluripotency, and function were maintained.
Shi et al., 2017 [37]	Extrusion	In vitro and in vivo	Silk fibroin and gelatin	MSCs	• Superior performance for cartilage repair in a knee joint as biomaterial matches mechanical properties of the newly formed cartilage.
Apelgren et al., 2017 [54]	Extrusion	In vivo	Nanofibrillated cellulose and alginate	Human chondrocytes and human MSCs	• Chondrocytes showed good proliferation ability. • In constructs comprising a mixture of chondrocytes and stem cells, an additional proliferative effect was observed involving chondrocyte production of glycosaminoglycans and type 2 collagen.

tures or equivalent bench-based printed scaffolds. A higher amount of newly regenerated cartilage was noted with the absence of subchondral deformation or collapse.

Although the aforementioned studies proposed that the use of cartilage 3D bioprinting for the treatment of focal defects will be feasible soon, it is unclear whether a holistic approach will be required when the whole joint is affected by osteoarthritis. This argument is supported by current evidence suggesting that osteoarthritis results in extensive changes, which are not limited to the cartilage but involve the entire joint including the subchondral bone [43]. Hence, a number of researchers focus on the production of osteochondral constructs rather than cartilage patches [44–48]. Woodfield et al. suggested that anatomically shaped, 3D bioprinted constructs with designed mechanical properties might offer alternatives for the reconstruction or restoration of congruent articulating surfaces [45]. In their study, 3D constructs loaded with chondrocytes were evaluated in vitro and in vivo (in rabbits). Fully functional chondrocytes were observed and the integration of the constructs with the bone was seen. Weight-bearing and functional joints were noted. In a similar study, a rabbit proximal humeral joint was captured with laser scanning and a scaffold was bioprinted layer-by-layer using HAp powder and PCL [46]. This scaffold was infused with transforming growth factor β3 (TGFβ3). The investigators reported that TGFβ3-infused bioscaffolds were fully covered with hyaline cartilage in their articular surface [46]. Similar approaches were also used by other researchers in animal models of femoral head and temporomandibular defects [47, 48].

3.4 Skin

Human skin is a complex structure having a variety of layers and cellular components. Skin loss from trauma and burns has been one of the earliest motivations of tissue engineering. Despite significant advances in skin tissue engineering, the designs often simplify considerably the structure of the skin to two main components (dermis and epidermis). Alternatively, 3D bioprinting has the potential to produce structures of higher complexity [Table 3.3] [55–67].

Three-dimensional bioprinting of human skin involved mainly loading of fibroblasts or/and keratinocytes into hydrogels of collagen, gelatin, or alginate [55]. The results of many studies showed that the resulting 3D engineered skin achieved high cellular survival and its histological appearance resembles that of human skin. In a slight deviation of most of the available studies, Koch et al. has added MSCs to the bioink [59]. They reported that MSCs retained high survival during the printing process and did not become apoptotic following the construction of the graft. The aforementioned approaches often result in low stability of the construct; hence crosslinking is required. To overcome this drawback, Min et al. printed fibroblasts, melanocytes, and keratinocytes onto collagen hydrogel cross-linked through neutralization using sodium bicarbonate [63]. The authors reported that the resulting melanocyte-containing epidermal layer showed freckle-like pig-

Table 3.3 Selected studies showing the current evidence on 3D bioprinting of skin

Author, Year	Printer	Design	Scaffold/Bioink	Cells	Outcome
Lee et al., 2014 [55]	Extrusion	In vitro	Collagen type I	Keratinocytes and fibroblasts	• 3D printed skin tissue had histological similarities to the human skin tissue
Lee et al., 2009 [58]	Extrusion	In vitro	Multilayer hydrogel	Keratinocytes and fibroblasts	• Highly viable proliferation of each cell layer was observed. • Organo-typic skin tissue culture is feasible.
Binder et al., 2010 [65]	In situ skin printer	In vivo (mice)		Keratinocytes and fibroblasts	• Acceptable survival rate of cells after printing was noted. • Fast healing rate of the skin defects occurred.
Koch et al., 2010 [60]	Laser-induced forward transfer	In vitro	Alginate hydrogel	Keratinocytes and fibroblasts and human MSCs	• High cells' survival of the printing process. • All used cell types maintained their ability to proliferate after printing. • Skin cells and hMSC showed no increase of apoptosis or DNA fragmentation.
Skardal et al., 2010 [66]	In situ extrusion	In vivo (mice)	Fibrin-collagen gel	Amniotic fluid cells and bone marrow-derived MSCs	• The graft resulted in higher re-epithelialization with increased microvessel density and capillary diameters. • The secreted trophic factors could be responsible for the favourable effect, rather than direct cell–cell interactions.
Albanna et al., 2012 [67]	In-situ extrusion	In vivo (porcine)	Fibrogen/ collagen solution	Fibroblasts and keratinocytes	• In situ skin bioprinting is a viable option for treatment of large skin defects. • The utilization of autologous cells outperformed in healing potential compared to allogeneic cell use.

(continued)

Table 3.3 (continued)

Author, Year	Printer	Design	Scaffold/Bioink	Cells	Outcome
Koch et al., 2012 [59]	Laser-assisted	In vitro	Collagen type I	Keratinocytes and fibroblasts	• Laser-assisted bioprinting is an outstanding tool for the generation of multicellular 3D resampling human skin.
Michael et al., 2013 [61]	Laser-assisted	In vitro and in vivo (mice)	Collagen type I	Keratinocytes and fibroblasts	• In vitro experiments showed proliferative cells, but they were in the whole epidermis. • Printed fibroblasts produced collagen. • In the mice, some blood vessels could be found to grow from the wound bed and the wound edges in the direction of the printed cells.
Cubo et al., 2016 [56]	Extrusion	In vitro and in vivo (mice)	Bioinks containing human plasma fibrin	Keratinocytes and fibroblasts	• Generated skin was similar to human skin. • Skin was indistinguishable from bilayered dermo-epidermal equivalents.
Liu et al., 2016 [64]	n/a	In vivo (mice)	Gelatin-alginate scaffold	No cells used	• 3D printed scaffold accelerated wound healing.
Kim et al., 2017 [62]	Inject and extrusion	In vitro	Gelatin and collagen I	Keratinocytes and fibroblasts	• Keratinocytes were uniformly distributed into the engineered dermis. • Maturation of a skin occurred. • Favourable biological characteristics including a stabilized fibroblast-stretched dermis and stratified epidermis were noted.

(continued)

Table 3.3 (continued)

Author, Year	Printer	Design	Scaffold/Bioink	Cells	Outcome
Pourchet et al., 2017 [57]	Extrusion	In vitro	Gelatin (Bovine)	Keratinocytes and fibroblasts	• Immuno-staining and electronic microscopy presented all characteristics of human skin. • The printability of large skin objects is demonstrated with the printing of an adult-size ear.
Min et al., 2018 [63]	Extrusion	In vitro	Collagen hydrogel crosslinked through neutralization using sodium bicarbonate	Fibroblasts, melanocytes, and keratinocytes	• Melanocyte containing epidermal layer showed freckle-like pigmentations at the dermal-epidermal junction, without the use of external ultraviolet light or chemical stimuli.

mentations at the dermal-epidermal junction, without external ultraviolet light or chemical stimuli [63].

A different approach to the aforementioned studies was reported by three different investigators exploring on-demand in situ 3D bioprinting [65–67]. In the treatment of a full-thickness skin defect model, Binder et al., 3D applied printed constructs containing keratinocytes and fibroblasts [65]. The resulting skin was similar to normal skin and complete wound healing was reported [65]. Skardal et al. used a full-thickness skin wound model where 3D printed amniotic fluid cells and bone marrow-derived MSCs suspended in fibrin-collagen gel were placed on the defects [66]. This approach resulted in higher levels of re-epithelialization and increased microvessel density and capillary diameters. Due to the fact that the printed cells did not permanently integrate with the surrounding tissues, authors concluded that the secreted trophic factors could be responsible for the favourable effect, rather than direct cell–cell interactions. In the third study, an experimentally induced 10x10cm skin defect in a porcine model was created [67]. The investigators explored the healing potential of fibroblasts and keratinocytes suspended in fibrogen/collagen solution and compared the overall potential of autologous versus allogeneic cells [67]. The results showed that this technique is a viable option for the treatment of large skin defects with the autologous cells outperform the use of allogeneic cells in terms of healing potential.

3.5 Neural Tissues

At present, the ideal approach for nerve repair is the precise microsurgical implantation of a healthy autologous nerve graft. This is the closest resemblance to the original microstructure of the missing nerve [68]. Although this is our current gold standard approach, the technique is associated with poor nerve function, donor site morbidity, and the formation of neuromas [68]. It has been proposed that 3D bioprinting could offer great potential in fabricating the precise cellular structures for nerve tissues.

For clinical scenarios where nerve transection occurs without nerve loss, the application of a 3D bioprinted fibrin scaffold created by extruding fibrinogen solution into thrombin solution and utilizing HA and polyvinyl alcohol, was found to mimic the natural fibrin clot that forms between injured nerve ends and encapsulated Schwann cells, thus providing natural guidance of neurite growth [69]. In cases of nerve damage with loss of neural tissue, approaches of 3D bioprinting could be divided in those aiming to construct hollow nerve guidance conduits or constructing more complex tissues with cells within complex bioinks. Three-dimensional printed hollow nerve conduits can be of natural or synthetic materials, single lumen or multilumen [70–72]. In vivo experiments have shown that these materials could promote nerve regeneration [70–72]. Alternatively, more complex 3D printed constructs using cells have also been found to promote nerve regeneration in experimental animal studies [73–75]. In particular, using 3D bioprinted scaffold-free conduits made from human normal dermal fibroblasts in an experimental animal model of transacted sciatic nerve, Yurie et al. reported favourable outcomes in the regeneration of the nerve [74]. Similar results were reported in an experimental tibial nerve injury model in rats [75]. Bioprinted cryopolymerized GelMA (cryoGelMA) gel was cellularized with adipose-derived stem cells. This graft could support the re-innervation across a 10 mm sciatic nerve gap in rats, with results close to those obtained with the use of autografts in terms of functional and histological characteristics [73]. In a different approach, a 3D printed layer-by-layer cylindrical structure loaded with cell suspension composed of 90% MSCs and 10% Schwann cells was used in the treatment of experimental sciatic nerve defects in rats [74]. The investigators in this concept study reported favourable results, recognizing the importance of several adjustments that need to be made. These adjustments include the removal of agarose rods from the construct lumina prior to implantation or using a hydrogel with faster degradation time in vivo, adjusting the number of lumina and modifying the cell types used or adding growth factors [74].

3.6 Blood Vessels

One of the major challenges in tissue engineering is the fabrication of vasculature or vascularized tissue. It has been previously noted that cells can survive at a distance of 200–400 μm from a blood vessel as the farthest [76]. To overcome the lack of vasculature, tissue engineering approaches have employed the addition of angiogenic factors such as the vascular endothelial growth factor (VEGF) to promote vascular migration from the host or employing surgical vascularized flaps [12, 77].

Several studies have evaluated the construction of vasculature through 3D bioprinting technologies [78–87]. Large blood vessels such as aortic tissue construct was 3D bioprinted layer-by-layer using mouse embryonic fibroblast cell aggregates and hydrogels. Smaller blood vessels were also constructed using 3D bioprinting. Tubular structures with 300 μm wall thickness, inner diameters of 1–2 mm, and defined pores with a constant diameter of approximately 100 or 200 μm mimicking the structure of blood vessels were also 3D printed [83]. In the work of Zhao et al., robotic 3D cell printing technology with a mesoscopic fluorescence molecular tomography imaging system was used to construct perfused collagen scaffolds with endothelial lining [78]. The authors imaged both the fluid flow and fluorescent-labelled living endothelial cells at high rates, with high sensitivity and accuracy [78] Finally, a more sophisticated construct was presented by the Atala group [87]. The authors documented the development of an integrated tissue-organ printer that can produce human-scale tissue constructs of various shapes and incorporating microchannels that allow for the diffusion of nutrients to printed cells. These tissues could be sustained for long periods, enabling the differentiation of cells into various lineages [82].

3.7 Muscle

Injuries to the skeletal muscles are debilitating and they result in extensive scaring which leads to functional impairment. Advances in 3D bioprinting showed significant potential for application in muscle regeneration [6, 88–91].

Early studies of 3D bioprinted myoblasts onto micro-sized cantilevers showed fusion of myoblasts to mature myotubes in 4 days of culture [89]. Alternatively, 3D printing of fibronectin stripes onto biodegradable L-lactide/trimethylene carbonate copolymer (PLLA-TMC) films with murine myoblast induced cell alignment and improved myotube formation [90]. In another study by Peele et al., muscle mimicking the function of musculature was printed using layer-by-layer stereolithography technique at high resolutions of 37 μm [91]. Merceron et al. have 3D printed a muscle-tendon unit resembling a functional human muscle [88]. This construct was developed in two layers. The first layer was composed of thermoplastic polyurethane co-printed with C2C12 cell-laden hydrogel-based bioink for elasticity and muscle development. The other layer was composed of PCL co-printed with

NIH/3T3 cell-laden hydrogel-based bioink for stiffness and tendon development on the other [88]. It exhibited high cell viability and allowed cellular differentiation [88]. Finally, muscle tissue that can respond to electrical stimulation in vivo was created by Kang et al. [87]. In this study, skeletal muscle constructs with the size of 15 mm × 5 mm × 1 mm were created by printing cell-laden hydrogels with biodegradable polymers in integrated patterns and anchored on sacrificial hydrogels.

3.8 Cardiovascular Tissue

Cardiovascular disease is one of the main causes of death worldwide [91, 92]. Tissue engineering approaches have focused on the regeneration of myocardium and the replacement of cardiovascular structures such as heart halves [93, 94]. The main aim of this technology is to fabricate biocompatible and non-immunogenic cardiac tissues having morphological and functional properties of the human heart.

In myocardial regeneration, 3D printed patch fabricated by using nano-reinforced hybrid cardiac patch laden with human coronary artery endothelial cells, methacrylated collagen micropatterning, and an alginate matrix was found to allow significant cellular proliferation, migration, and differentiation [95]. In a similar study, the fabrication of a cardiac patch composed of human cardiac-derived progenitor cells (hCMPCs) in a HA/gelatin (HA/gel) based matrix lead to the preservation of cardiac performance in myocardial infarction model in mice [96] In another study, cell-laden hydrogel printed with a sacrificial hydrogel resulted in the formation of cardiac tissue constructs that exhibited spontaneous synchronous contraction in culture. This implies in vitro cardiac tissue development and maturation [97].

In heart valve repair, literature has shown that 3D bioprinting technology is a promising tool in constructing valves to meet the biomechanical and haemodynamic requirements [98–100]. Hockaday et al. presented a novel simultaneous 3D printing/photo-crosslinking technique for rapidly engineering complex, heterogeneous aortic valve scaffolds [98]. The investigators proposed that these constructs can be fabricated rapidly and when they were seeded with porcine aortic valve interstitial cells, these cells maintained a nearly 100% viability over 21 days of culture [98]. High cellular viability was also reported in a similar study of 3D bioprinted alginate/gelatin hydrogel valve conduits with anatomical architecture and direct incorporation of aortic root sinus smooth muscle cells (SMC) and aortic valve leaflet interstitial cells in a regionally constrained manner [99].

3.9 Other Tissues

Three-dimensional bioprinting has found applications in several other fields of regenerative medicine. Human tissues such as liver, trachea, and retina were also created. Printing of constructs that resemble human liver and allow heterotypic cellular interactions within the resulting structures was proposed [101, 102] Three-dimensionally printed PCL mimicking human trachea was also reported [103]. The investigators highlighted the importance of cells in this technique, as severe inflammation and an unorganized structure occurred when it was implanted in rabbits [103]. The latter complication was reduced significantly when the graft was cultured in the omentum for 2 weeks [102]. Other applications of 3D printing include also the construction of human ear or auricular cartilage, meniscal tissues, and other tissue analogues [104–107]. Extensive research is currently underway exploring the feasibility of fabricating of human retina [108–111]. A 3D printing of retinal and glial cells as a retina model was demonstrated by Lorber et al. [108]. The printed cells seemed to retain their growth-promoting properties and their viability, unaffected by the piezoelectric printhead [108] The differentiation of retinal cells seemed influenced by the extracellular matrix [109] More specifically, it seems crucial to recapitulate the extracellular environment of these cells, so it can mimic the stiffness of the human retina, which seems to promote cell differentiation [108, 110, 112]. To this end, 3D-bioprinting of HA hydrogels with the addition of retinal progenitor cells was found to have favourable results [110].

The fabrication of pathological tissue models for research is also possible using 3D bioprinting. An in vitro cervical tumour model using Hela cells and gelatin/alginate/fibrinogen hydrogels was constructed [113]. Zhou et al. developed a biomimetic bone matrix using 3D bioprinting technology to investigate the interaction between breast cancer cells and fetal osteoblasts or human bone marrow MSCs [114]. The authors suggested that this was a suitable model to study the interactive effects of cells in the context of an artificial bone microenvironment and may thus serve as a valuable tool for the investigation of post-metastatic breast cancer progression in bone [114]. Finally, the development of a perfusable vascularized 3D tissue resembling liver tissue was used to study drug toxicity in vitro [115].

3.10 Conclusions

Three-dimensional bioprinting has evolved rapidly over the last decade as a promising tool in tissue regeneration. Its main advantages are the high precision of tissue fabrication and a fast construction speed. At present, there is an abundance of studies showing the potential of this technology in vitro and in several animal models. It is indisputable that in comparison with other tissue engineering approaches, 3D bioprinting holds most ground as it enables the fabrication of biomimetic tissues. Several challenges can be identified including the biomechanical control, the selection of scaffolds, and the safeguarding of safety throughout the process until the implantation of the constructs into the patient takes place. Other challenges include the vascularization of the 3D printed constructs and the overall survival in the body. These challenges will hopefully be overcome soon through collaborations between medics, biologists, bioengineers, and physicians.

Acknowledgements *Conflict of Interest*: No benefits in any form have been received or will be received from a commercial party related directly or indirectly to the subject of this chapter.

References

1. Ashammakhi N, Kaarela O, Hasan A, Byambaa B, Sheikhi A, Gaharwar AK, Khademhosseini A (2019) Advancing frontiers in bone bioprinting. Adv Healthc Mater 8:e1801048
2. Tellisi N, Ashammakhi NA, Billi F, Kaarela O (2018) Three dimensional printed bone implants in the clinic. J Craniofac Surg 29:2363–2367
3. Skardal A, Atala A (2015) Biomaterials for integration with 3-D bioprinting. Ann Biomed Eng 43:730–746
4. Lei M, Wang X (2018) Biodegradable polymers and stem cells for bioprinting. Molecules 21(5):pii: E539
5. Malda J, Visser J, Melchels FP, Jüngst T, Hennink WE, Dhert WJ, Groll J, Hutmacher DW (2013) 25th anniversary article: engineering hydrogels for biofabrication. Adv Mater 25:5011–5028
6. Jessop ZM, Al-Sabah A, Gardiner MD, Combellack E, Hawkins K, Whitaker IS (2007) 3D bioprinting for reconstructive surgery: principles, applications and challenges. J Plast Reconstr Aesthet Surg 70(9):1155–1170
7. Ashammakhi N, Ahadian S, Pountos I, Hu S-K, Tellisi N, Bandaru P, Ostrovidov S, Dokmeci M, Khademhosseini A (2019) In situ three-dimensional printing for reparative and regenerative therapy. Biomed Microdevices 21(2):42
8. Hong N, Yang GH, Lee J, Kim G (2018) 3D bioprinting and its in vivo applications. J Biomed Mater Res B Appl Biomater 106:444–459
9. Luo Y, Lin X, Huang P (2018) 3D Bioprinting of artificial tissues: construction of biomimetic microstructures. Macromol Biosci 18:e1800034
10. Pountos I, Giannoudis PV (2016) Is there a role of coral bone substitutes in bone repair? Injury 47:2606–2613
11. Panteli M, Pountos I, Jones E, Giannoudis PV (2015) Biological and molecular profile of fracture non-union tissue: current insights. J Cell Mol Med 19:685–713
12. Pountos I, Panteli M, Panagiotopoulos E, Jones E, Giannoudis PV (2014) Can we enhance fracture vascularity: what is the evidence? Injury 45(Suppl 2):S49–S57
13. Pountos I, Georgouli T, Pneumaticos S, Giannoudis PV (2013) Fracture non-union: can biomarkers predict outcome? Injury 44:1725–1732
14. Bajada S, Harrison PE, Ashton BA, Cassar-Pullicino VN, Ashammakhi N, Richardson JB (2007) Successful treatment of refractory tibial nonunion using calcium sulphate and bone marrow stromal cell implantation. J Bone Joint Surg Br 89:1382–1386
15. Poldervaart MT, Goversen B, de Ruijter M, Abbadessa A, Melchels FPW, Öner FC, Dhert WJA, Vermonden T, Alblas J (2017) 3D bioprinting of methacrylated hyaluronic acid (MeHA) hydrogel with intrinsic osteogenicity. PLoS One 12:e0177628
16. Wang XF, Song Y, Liu YS, Sun YC, Wang YG, Wang Y, Lyu PJ (2016) Osteogenic differentiation of three-dimensional bioprinted constructs consisting of human adipose-derived stem cells in vitro and in vivo. PLoS One 11:e0157214
17. Poldervaart MT, Wang H, van der Stok J, Weinans H, Leeuwenburgh SC, Öner FC, Dhert WJ, Alblas J (2013) Sustained release of BMP-2 in bioprinted alginate for osteogenicity in mice and rats. PLoS One 8(8):e72610
18. Duarte Campos DF, Blaeser A, Buellesbach K, Sen KS, Xun W, Tillmann W, Fischer H (2016) Bioprinting organotypic hydrogels with improved mesenchymal stem cell remodeling and mineralization properties for bone tissue engineering. Adv Healthc Mater 5:1336–1345

19. Seyednejad H, Gawlitta D, Kuiper RV, de Brui A, van Nostrum CF, Vermonden T, Dhert WJA, Hennink WE (2012) In vivo biocompatibility and biodegradation of 3D-printed porous scaffolds based on a hydroxyl-functionalized poly(ε-caprolactone). Biomaterials 33:4309
20. Park SH, Park DS, Shin JW, Kang YG, Kim HK, Yoon TR, Shin JW (2012) Scaffolds for bone tissue engineering fabricated from two different materials by the rapid prototyping technique: PCL versus PLGA. J Mater Sci Mater Med 23:2671
21. Heo SJ, Kim SE, Wei J, Kim DH, Hyun YT, Yun HS, Kim HK, Yoon TR, Kim SH, Park SA, Shin JW, Shin JW (2009) In vitro and animal study of novel nano-hydroxyapatite/poly(epsilon-caprolactone) composite scaffolds fabricated by layer manufacturing process. Tissue Eng Part A 15:977–989
22. Kim J, McBride S, Brandi T, Alvarez-Urena P, Song YH, Dean DD, Sylvia VL, Elgendy H, Ong J, Hollinger JO (2012) Rapid-prototyped PLGA/β-TCP/hydroxyapatite nanocomposite scaffolds in a rabbit femoral defect model. Biofabrication 4(2):025003
23. Gao G, Schilling AF, Yonezawa T, Wang J, Dai G, Cui X (2014) Bioactive nanoparticles stimulate bone tissue formation in bioprinted three-dimensional scaffold and human mesenchymal stem cells. Biotechnol J 9:1304–1311
24. Martin Y, Eldardiri M, Lawrence-Watt DJ, Sharpe JR (2011) Microcarriers and their potential in tissue regeneration. Tissue Eng Part B Rev 17:71–80
25. Sart S, Agathos SN, Li Y (2013) Engineering stem cell fate with biochemical and biomechanical properties of microcarriers. Biotechnol Prog 29:1354–1366
26. Luo Y, Wu C, Lode A, Gelinsky M (2013) Hierarchical mesoporous bioactive glass/alginate composite scaffolds fabricated by three-dimensional plotting for bone tissue engineering. Biofabrication 5:15005
27. Reichert JC, Wullschleger ME, Cipitria A, Lienau J, Cheng TK, Schütz MA, Duda GN, Nöth U, Eulert J, Hutmacher DW (2011) Custom-made composite scaffolds for segmental defect repair in long bones. Int Orthop 35:1229–1236
28. Keriquel V, Guillemot F, Arnault I, Guillotin B, Miraux S, Amédée J, Fricain JC, Catros S (2010) In vivo bioprinting for computer- and robotic-assisted medical intervention: preliminary study in mice. Biofabrication 2:014101
29. Keriquel V, Oliveira H, Rémy M, Ziane S, Delmond S, Rousseau B, Rey S, Catros S, Amédée J, Guillemot F, Fricain JC (2017) In situ printing of mesenchymal stromal cells, by laser-assisted bioprinting, for in vivo bone regeneration applications. Sci Rep 7:1778
30. Bendtsen ST, Quinnell SP, Wei M (2017) Development of a novel alginate-polyvinyl alcohol-hydroxyapatite hydrogel for 3D bioprinting bone tissue engineered scaffolds. J Biomed Mater Res A 105:1457–1468
31. Demirtaş TT, Irmak G, Gümüşderelioğlu M (2017) A bioprintable form of chitosan hydrogel for bone tissue engineering. Biofabrication 9:035003
32. Neufurth M, Wang X, Wang S, Steffen R, Ackermann M, Haep ND, Schröder HC, Müller WEG (2017) 3D printing of hybrid biomaterials for bone tissue engineering: calcium-polyphosphate microparticles encapsulated by polycaprolactone. Acta Biomater 64:377–388
33. Zhang Y, Zhai D, Xu M, Yao Q, Zhu H, Chang J, Wu C (2017) 3D-printed bioceramic scaffolds with antibacterial and osteogenic activity. Biofabrication 9:025037
34. Du M, Chen B, Meng Q, Liu S, Zheng X, Zhang C, Wang H, Li H, Wang N, Dai J (2015) 3D bioprinting of BMSC-laden methacrylamide gelatin scaffolds with CBD-BMP2-collagen microfibers. Biofabrication 7:044104
35. Buckwalter JA, Saltzman C, Brown T (2004) The impact of osteoarthritis: implications for research. Clin Orthop Relat Res (427 Suppl):S6–S15
36. Ashammakhi N, Ahadian S, Darabi MA, Tahchi ME, Lee J, Suthiwanich K, Sheikhi A, Dokmeci MR, Oklu R, Khademhosseini A (2018) Minimally invasive and regenerative therapeutics. Adv Mater 22:e1804041
37. Shi W, Sun M, Hu X, Ren B, Cheng J, Li C, Duan X, Fu X, Zhang J, Chen H, Ao Y (2017) Structurally and functionally optimized silk-fibroin-gelatin scaffold using 3D printing to repair cartilage injury in vitro and in vivo. Adv Mater 29(29)

38. Cui X, Breitenkamp K, Finn MG, Lotz M, D'Lima DD (2012) Direct human cartilage repair using three-dimensional bioprinting technology. Tissue Eng Part A 18:1304–1312
39. Kundu J, Shim JH, Jang J, Kim SW, Cho DW (2015) An additive manufacturing-based PCL-alginate-chondrocyte bioprinted scaffold for cartilage tissue engineering. J Tissue Eng Regen Med 9:1286–1297
40. Xu T, Binder KW, Albanna MZ, Dice D, Zhao W, Yoo JJ, Atala A (2013) Hybrid printing of mechanically and biologically improved constructs for cartilage tissue engineering applications. Biofabrication 5:015001
41. Di Bella C, Duchi S, O'Connell CD, Blanchard R, Augustine C, Yue Z, Thompson F, Richards C, Beirne S, Onofrillo C, Bauquier SH, Ryan SD, Pivonka P, Wallace GG, Choong PF (2018) In situ handheld three-dimensional bioprinting for cartilage regeneration. J Tissue Eng Regen Med 12(3):611–621
42. Daly AC, Critchley SE, Rencsok EM, Kelly DJ (2016) A comparison of different bioinks for 3D bioprinting of fibrocartilage and hyaline cartilage. Biofabrication 8:045002
43. Pountos I, Giannoudis PV (2017) Modulation of cartilage's response to injury: can chondrocyte apoptosis be reversed? Injury 48:2657–2669
44. Shim JH, Jang KM, Hahn SK, Park JY, Jung H, Oh K, Park KM, Yeom J, Park SH, Kim SW, Wang JH, Kim K, Cho DW (2017) Three-dimensional bioprinting of multilayered constructs containing human mesenchymal stromal cells for osteochondral tissue regeneration in the rabbit knee joint. Biofabrication 8(1):014102
45. Woodfield TB, Guggenheim M, von Rechenberg B, Riesle J, van Blitterswijk CA, Wedler V (2009) Rapid prototyping of anatomically shaped, tissue-engineered implants for restoring congruent articulating surfaces in small joints. Cell Prolif 42:485–497
46. Lee CH, Cook JL, Mendelson A, Moioli EK, Yao H, Mao JJ (2010) Regeneration of the articular surface of the rabbit synovial joint by cell homing: a proof of concept study. Lancet 376:440–448
47. Tarafder S, Koch A, Jun Y, Chou C, Awadallah MR, Lee CH (2016) Micro-precise spatiotemporal delivery system embedded in 3D printing for complex tissue regeneration. Biofabrication 8:025003
48. Ding C, Qiao Z, Jiang W, Li H, Wei J, Zhou G, Dai K (2013) Regeneration of a goat femoral head using a tissue-specific, biphasic scaffold fabricated with CAD/CAM technology. Biomaterials 34:6706–6716
49. Schuurman W, Levett PA, Pot MW, van Weeren PR, Dhert WJ, Hutmacher DW, Melchels FP, Klein TJ, Malda J (2013) Gelatin-methacrylamide hydrogels as potential biomaterials for fabrication of tissue-engineered cartilage constructs. Macromol Biosci 13:551–561
50. Markstedt K, Mantas A, Tournier I, Martínez Ávila H, Hägg D, Gatenholm P (2015) 3D bioprinting human chondrocytes with nanocellulose-alginate bioink for cartilage tissue engineering applications. Biomacromolecules 16(5):1489–1496
51. Costantini M, Idaszek J, Szöke K, Jaroszewicz J, Dentini M, Barbetta A, Brinchmann JE, Święszkowski W (2016) 3D bioprinting of BM-MSCs-loaded ECM biomimetic hydrogels for in vitro neocartilage formation. Biofabrication 8(3):035002
52. Ren X, Wang F, Chen C, Gong X, Yin L, Yang L (2016) Engineering zonal cartilage through bioprinting collagen type II hydrogel constructs with biomimetic chondrocyte density gradient. BMC Musculoskelet Disord 17:301
53. Nguyen D, Hägg DA, Forsman A, Ekholm J, Nimkingratana P, Brantsing C, Kalogeropoulos T, Zaunz S, Concaro S, Brittberg M, Lindahl A, Gatenholm P, Enejder A, Simonsson S (2017) Cartilage tissue engineering by the 3D bioprinting of iPS cells in a nanocellulose/alginate bioink. Sci Rep 7:658
54. Apelgren P, Amoroso M, Lindahl A, Brantsing C, Rotter N, Gatenholm P, Kölby L (2017) Chondrocytes and stem cells in 3D-bioprinted structures create human cartilage in vivo. PLoS One 12:e0189428

55. Lee V, Singh G, Trasatti JP, Bjornsson C, Xu X, Tran TN, Yoo SS, Dai G, Karande P (2014) Design and fabrication of human skin by three-dimensional bioprinting. Tissue Eng Part C Methods 20:473–484
56. Cubo N, Garcia M, Del Cañizo JF, Velasco D, Jorcano JL (2016) 3D bioprinting of functional human skin: production and in vivo analysis. Biofabrication 9:015006
57. Pourchet LJ, Thepot A, Albouy M, Courtial EJ, Boher A, Blum LJ, Marquette CA (2017) Human skin 3D bioprinting using scaffold-free approach. Adv Healthc Mater 6(4):1601101
58. Lee W, Debasitis JC, Lee VK, Lee JH, Fischer K, Edminster K, Park JK, Yoo SS (2009) Multi-layered culture of human skin fibroblasts and keratinocytes through three-dimensional free form fabrication. Biomaterials 30:1587–1595
59. Koch L, Deiwick A, Schlie S, Michael S, Gruene M, Coger V, Zychlinski D, Schambach A, Reimers K, Vogt PM, Chichkov B (2012) Skin tissue generation by laser cell printing. Biotechnol Bioeng 109:1855–1863
60. Koch L, Kuhn S, Sorg H, Gruene M, Schlie S, Gaebel R, Polchow B, Reimers K, Stoelting S, Ma N, Vogt PM, Steinhoff G, Chichkov B (2010) Laser printing of skin cells and human stem cells. Tissue Eng Part C Methods 16:847–854
61. Michael S, Sorg H, Peck CT, Koch L, Deiwick A, Chichkov B, Vogt PM, Reimers K (2013) Tissue engineered skin substitutes created by laser-assisted bioprinting form skin-like structures in the dorsal skin fold chamber in mice. PLoS One 8:e57741
62. Kim BS, Lee JS, Gao G, Cho DW (2017) Direct 3D cell-printing of human skin with functional transwell system. Biofabrication 9(2):025034
63. Min D, Lee W, Bae IH, Lee TR, Croce P, Yoo SS (2018) Bioprinting of biomimetic skin containing melanocytes. Exp Dermatol 27:453–459
64. Liu J, Chi J, Wang K, Liu X, Liu J, Gu F (2016) Full-thickness wound healing using 3D bioprinted gelatin-alginate scaffolds in mice: a histopathological study. Int J Clin Exp Pathol 9(11):11197–11205
65. Binder KW, Zhao W, Aboushwareb T, Dice D, Atala A, Yoo JJ (2010) In situ bioprinting of the skin for burns. J Am Coll Surg 211:S76
66. Skardal A, Zhang J, McCoard L, Xu X, Oottamasathien S, Prestwich GD (2010) Photocrosslinkable hyaluronan-gelatin hydrogels for two-step bioprinting. Tissue Eng Part A 16:2675–2685
67. Albanna M, Murphy S, Zhao W, El-Amin I, Josh Tan J, Dennis Dice D, Kang HW, Jackson J, Atala A, Yoo J (2012) In situ bioprinting of skin for reconstruction. J Urol 187:e8
68. Pfister BJ, Gordon T, Loverde JR, Kochar AS, Mackinnon SE, Cullen DK (2011) Biomedical engineering strategies for peripheral nerve repair: surgical applications, state of the art, and future challenges. Crit Rev Biomed Eng 39:81–124
69. England S, Rajaram A, Schreyer DJ, Chen X (2017) Bioprinted fibrin-factor XIII-hyaluronate hydrogel scaffolds with encapsulated Schwann cells and their in vitro characterization for use in nerve regeneration. Bioprinting 5:1–9
70. Evans GR, Brandt K, Widmer MS, Lu L, Meszlenyi RK, Gupta PK, Mikos AG, Hodges J, Williams J, Gürlek A, Nabawi A, Lohman R, Patrick CW Jr (1999) In vivo evaluation of poly(L-lactic acid) porous conduits for peripheral nerve regeneration. Biomaterials 20:1109–1115
71. Radulescu D, Dhar S, Young CM, Taylor DM, Trost HJ, Hayes DJ, Evans GR (2007) Tissue engineering scaffolds for nerve regeneration manufactured by ink-jet technology. Mater Sci Eng C 23:534–539
72. Lee SJ, Zhu W, Heyburn L, Nowicki M, Harris B, Zhang LG (2017) Development of novel 3-D printed scaffolds with core-shell nanoparticles for nerve regeneration. IEEE Trans Biomed Eng 64:408–418
73. Hu Y, Wu Y, Gou Z, Tao J, Zhang J, Liu Q, Kang T, Jiang S, Huang S, He J, Chen S, Du Y, Gou M (2016) 3D-engineering of cellularized conduits for peripheral nerve regeneration. Sci Rep 6:32184

74. Yurie H, Ikeguchi R, Aoyama T, Kaizawa Y, Tajino J, Ito A, Ohta S, Oda H, Takeuchi H, Akieda S, Tsuji M, Nakayama K, Matsuda S (2017) The efficacy of a scaffold-free Bio 3D conduit developed from human fibroblasts on peripheral nerve regeneration in a rat sciatic nerve model. PLoS One 12:e0171448
75. Adams AM, VanDusen KW, Kostrominova TY, Mertens JP, Larkin LM (2017) Scaffoldless tissue-engineered nerve conduit promotes peripheral nerve regeneration and functional recovery after tibial nerve injury in rats. Neural Regen Res 12:1529–1537
76. Atala A, Kasper FK, Mikos AG (2012) Engineering complex tissues. Sci Transl Med 4:160rv12
77. Penttilä H, Tulamo R-M, Waris T, Ellä V, Kellomäki M, Törmälä P, Ashammakhi N (2004) Combining prefabricated microvascularied perichondrial flaps and bioabsorbable poly-lactide nonwoven scaffolds to tissue engineered cartilage. In: Joint Meeting of the Tissue Engineering Society International (TESI) and the European Tissue Engineering Society (ETES), Lausanne, Switzerland, 10–13.10.2004, P020
78. Zhao L, Lee VK, Yoo SS, Dai G, Intes X (2012) The integration of 3-D cell printing and mesoscopic fluorescence molecular tomography of vascular constructs within thick hydrogel scaffolds. Biomaterials 33:5325–5332
79. Wu W, DeConinck A, Lewis JA (2011) Omnidirectional printing of 3D microvascular networks. Adv Mater 23(24):H178–H183
80. Lee VK, Lanzi AM, Haygan N, Yoo SS, Vincent PA, Dai G (2014) Generation of multi-scale vascular network system within 3D hydrogel using 3D bio-printing technology. Cell Mol Bioeng 7(3):460–472
81. Kucukgul C, Ozler SB, Inci I, Karakas E, Irmak S, Gozuacik D, Taralp A, Koc B (2015) 3D bioprinting of biomimetic aortic vascular constructs with self-supporting cells. Biotechnol Bioeng 112(4):811–821
82. Kolesky DB, Homan KA, Skylar-Scott MA, Lewis JA (2016) Three-dimensional bioprinting of thick vascularized tissues. Proc Natl Acad Sci U S A 113(12):3179–3184
83. Huber B, Engelhardt S, Meyer W, Krüger H, Wenz A, Schönhaar V, Tovat GEM, Kluger PJ, Borchers K (2016) Blood-vessel mimicking structures by stereolithographic fabrication of small porous tubes using cytocompatible polyacrylate elastomers, biofunctionalization and endothelialization. J Funct Biomater 7(2):11
84. Kolesky DB, Truby RL, Gladman A, Busbee TA, Homan KA, Lewis JA (2014) 3D bioprinting of vascularized, heterogeneous cell-laden tissue constructs. Adv Mater 26(19):3124–3130
85. Meyer EP, Ulmann-Schuler A, Staufenbiel M, Krucker T (2008) Altered morphology and 3D architecture of brain vasculature in a mouse model for Alzheimer's disease. Proc Natl Acad Sci U S A 105(9):3587–3592
86. Xu Y, Hu Y, Liu C, Yao H, Liu B, Mi S (2018) A novel strategy for creating tissue-engineered biomimetic blood vessels using 3D bioprinting technology. Materials (Basel) 11(9):E1581
87. Kang HW, Lee SJ, Ko IK, Kengla C, Yoo JJ, Atala A (2016) A 3D bioprinting system to produce human-scale tissue constructs with structural integrity. Nat Biotechnol 34:312–319
88. Merceron TK, Burt M, Seol YJ, Kang HW, Lee SJ, Yoo JJ, Atala A (2015) A 3D bioprinted complex structure for engineering the muscle-tendon unit. Biofabrication 7:035003
89. Cui X, Gao G, Qiu Y (2013) Accelerated myotube formation using bioprinting technology for biosensor applications. Biotechnol Lett 35:315–321
90. Altomare L, Riehle M, Gadegaard N, Tanzi MC, Farè S (2010) Microcontact printing of fibronectin on a biodegradable polymeric surface for skeletal muscle cell orientation. Int J Artif Organs 33:535–543
91. Peele BN, Wallin TJ, Zhao H, Shepherd RF (2015) 3D printing antagonistic systems of artificial muscle using projection stereolithography. Bioinspir Biomim 10:055003
92. McAloon CJ, Boylan LM, Hamborg T, Stallard N, Osman F, Lim PB, Hayat SA (2016) The changing face of cardiovascular disease 2000-2012: an analysis of the world health organisation global health estimates data. Int J Cardiol 224:256–264

93. Smit FE, Dohmen PM (2015) Cardiovascular tissue engineering: where we come from and where are we now? Med Sci Monit Basic Res 21:1–3
94. Weinberger F, Mannhardt I, Eschenhagen T (2017) Engineering cardiac muscle tissue: a maturating field of research. Circ Res 120:1487–1500
95. Izadifar M, Chapman D, Babyn P, Chen X, Kelly ME (2018) UV-assisted 3D bioprinting of nanoreinforced hybrid cardiac patch for myocardial tissue engineering. Tissue Eng Part C Methods 24:74–88
96. Gaetani R, Feyen DA, Verhage V, Slaats R, Messina E, Christman KL, Giacomello A, Doevendans PA, Sluijter JP (2015) Epicardial application of cardiac progenitor cells in a 3D-printed gelatin/hyaluronic acid patch preserves cardiac function after myocardial infarction. Biomaterials 61:339–348
97. Wang Z, Lee SJ, Cheng HJ, Yoo JJ, Atala A (2018) 3D bioprinted functional and contractile cardiac tissue constructs. Acta Biomater 70:48–56
98. Hockaday LA, Kang KH, Colangelo NW, Cheung PY, Duan B, Malone E, Wu J, Girardi LN, Bonassar LJ, Lipson H, Chu CC, Butcher JT (2012) Rapid 3D printing of anatomically accurate and mechanically heterogeneous aortic valve hydrogelscaffolds. Biofabrication 4:035005
99. Duan B, Hockaday LA, Kang KH, Butcher JT (2013) 3D bioprinting of heterogeneous aortic valve conduits with alginate/gelatin hydrogels. J Biomed Mater Res A 101:1255–1264
100. Kang LH, Armstrong PA, Lee LJ, Duan B, Kang KH, Butcher JT (2017) Optimizing photo-encapsulation viability of heart valve cell types in 3D printable composite hydrogels. Ann Biomed Eng 45:360–377
101. Lee JW, Choi YJ, Yong WJ, Pati F, Shim JH, Kang KS, Kang IH, Park J, Cho DW (2016) Development of a 3D cell printed construct considering angiogenesis for liver tissue engineering. Biofabrication 8:015007
102. Chang R, Emami K, Wu H, Sun W (2010) Biofabrication of a three-dimensional liver micro-organ as an in vitro drug metabolism model. Biofabrication 2(4):045004
103. Park HS, Lee JS, Jung H, Kim DY, Kim SW, Sultan MT, Park CH (2018) An omentum-cultured 3D-printed artificial trachea: in vivo bioreactor. Artif Cells Nanomed Biotechnol 19:1–10
104. Ávila HM, Schwarz S, Rotter N, Gatenholm P (2016) 3D bioprinting of human chondrocyte-laden nanocellulose hydrogels for patient-specific auricular cartilage regeneration. Bioprinting 1–2:22–35
105. Lee JS, Hong JM, Jung JW, Shim JH, Oh JH, Cho DW (2014) 3D printing of composite tissue with complex shape applied to ear regeneration. Biofabrication 6(2):024103
106. Reiffel AJ, Kafka C, Hernandez KA, Popa S, Perez JL, Zhou S, Pramanik S, Brown BN, Ryu WS, Bonassar LJ, Spector JA (2013) High-fidelity tissue engineering of patient-specific auricles for reconstruction of pediatric microtia and other auricular deformities. PLoS One 8(2):e56506
107. Mannoor MS, Jiang Z, James T, Kong YL, Malatesta KA, Soboyejo WO, Verma N, Gracias DH, McAlpine MC (2013) 3D printed bionic ears. Nano Lett 13:2634–2639
108. Lorber B, Hsiao WK, Hutchings IM, Martin KR (2014) Adult rat retinal ganglion cells and glia can be printed by piezoelectric inkjet printing. Biofabrication 6:015001
109. Hunt NC, Hallam D, Karimi A, Mellough CM, Chen J, Steel DM, La M (2016) 3D culture of human pluripotent stem cells in alginate hydrogel improves retinal tissue development. Acta Biomater 49:329–343. https://doi.org/10.1017/CBO9781107415324.004
110. Wang P, Li X, Zhu W, Zhong Z, Moran A, Wang W, Zhang K, Chen S (2018) 3D bioprinting of hydrogels for retina cell culturing. Bioprinting 12:e00029
111. Shi P, Tan YSE, Yeong WY, Li HY, Laude A (2018) A bilayer photoreceptor-retinal tissue model with gradient cell density design: a study of microvalve-based bioprinting. J Tissue Eng Regen Med 12:1297–1306

112. Mitrousis N, Tam RY, Baker AEG, Van Der Kooy D, Shoichet MS (2016) Hyaluronic acid-based hydrogels enable rod photoreceptor survival and maturation in vitro through activation of the mTOR pathway. Adv Funct Mater 26:1975–1985
113. Zhao Y, Yao R, Ouyang L, Ding H, Zhang T, Zhang K, Cheng S, Sun W (2014) Three-dimensional printing of Hela cells for cervical tumor model in vitro. Biofabrication 6:035001
114. Zhou X, Zhu W, Nowicki M, Miao S, Cui H, Holmes B, Glazer RI, Zhang LG (2016) 3D bioprinting a cell-laden bone matrix for breast cancer metastasis study. ACS Appl Mater Interfaces 8:30017–30026
115. Massa S, Sakr MA, Seo J, Bandaru P, Arneri A, Bersini S, Zare-Eelanjegh E, Jalilian E, Cha BH, Antona S, Enrico A, Gao Y, Hassan S, Acevedo JP, Dokmeci MR, Zhang YS, Khademhosseini A, Shin SR (2017) Bioprinted 3D vascularized tissue model for drug toxicity analysis. Biomicrofluidics 11:044109

Chapter 4
Biopriting Vasculature

Sanskrita Das and Jinah Jang

Abstract Despite the extensive research in fabricating tissue-engineered vascularized constructs, emulating the native architecture with intricate microvascular networks in vitro remains challenging, which limits clinical applications. The 3D bioprinting technique is a promising approach for overcoming the limitations posed by the classical tissue engineering strategies. The new generation of bioprinted vascularized tissue constructs facilitates the high spatial control of cell allocation, alignment, and maturation and vessel stabilization as a result of the efficient diffusion of oxygen, nutrients, and (optionally) growth factors, thereby enhancing the metabolic activity of cells. Moreover, the bioprinted vascularized construct accelerates its integration with the host tissue upon implantation, promoting rapid microvascular formation and tissue regeneration. Additionally, the flexibility to fabricate cell-laden, multi-material, and anatomically shaped vascular grafts and vascularized tissue constructs encourages the development of modalities for screening new therapeutic drugs and for using as an in vitro disease model. In this chapter, we briefly discuss the need for using tissue-engineered vascularized constructs and summarize the different types of biomaterials and conventional approaches toward it. We also introduce the advent of 3D bioprinting in developing 3D vascularized constructs and focus on its applications in tissue regeneration and as a platform for drug discovery and testing.

Keywords 3D bioprinting · Bioink · Vascularization · Solid organ

S. Das
Department of Creative IT Engineering, Pohang University of Science and Technology, Pohang, Republic of Korea
e-mail: sanskrita@postech.ac.kr

J. Jang (✉)
Department of Creative IT Engineering, Pohang University of Science and Technology, Pohang, Republic of Korea

Department of Mechanical Engineering, Pohang University of Science and Technology, Pohang, Republic of Korea

School of Interdisciplinary Bioscience and Bioengineering, Pohang University of Science and Technology, Pohang, Republic of Korea
e-mail: jinahjang@postech.ac.kr

M. Guvendiren (ed.), *3D Bioprinting in Medicine*,
https://doi.org/10.1007/978-3-030-23906-0_4

4.1 Introduction

The human vascular system is a complex structure with a network of blood vessels ranging from the centimeter scale to the micrometer scale (i.e., from ~2.5 cm for the aorta to ~20 μm for very fine capillaries) [1]. The vascular system is largely involved in the exchange of nutrients, oxygen, and cellular metabolic waste between the blood and tissues [1, 2]. This is mediated through a monolayer of flat cells, called endothelial cells (ECs), which line the lumen of the entire vascular system. Moreover, this layer can maintain hemostasis and has anti-thrombogenic properties necessary for maintaining normal tissue function [1, 2].

The vascularization process can be divided into four phases starting from the (1) sprouting of ECs, (2) the anastomosis of the sprouts, (3) lumen formation, and (4) retraction of unused capillaries [2]. Briefly, the pre-ECs form the capillary sprouts, which further anastomose with their neighboring sprouts developing into a primitive capillary structure with multiple junction points. After anastomosis, the lumen is formed in each sprout but away from their junction point followed by the formation of a continuous lumen, also called a plexus. Finally, one of the sprouts enlarges as a major pathway for blood flow, whereas the remaining capillary sprouts are retracted as capillary blood vessels. The maximum distance between the capillaries is 200 μm, corresponding to the diffusion limit of oxygen [4] (Fig. 4.1).

Dysfunctional vasculature or damaged blood vessel conduits as a result of trauma, aging, or acute/chronic diseases is a common underlying cause for damaged tissue/organ function as the native vessels are inadequate to supply the required blood to the tissues of the heart, legs, and other organs [5, 6]. In the case of arterial bypass surgeries, autologous vessels, auxiliary arteries, and veins are the prevailing choices for bypass grafting [7, 8]. Nevertheless, these are associated with challenges

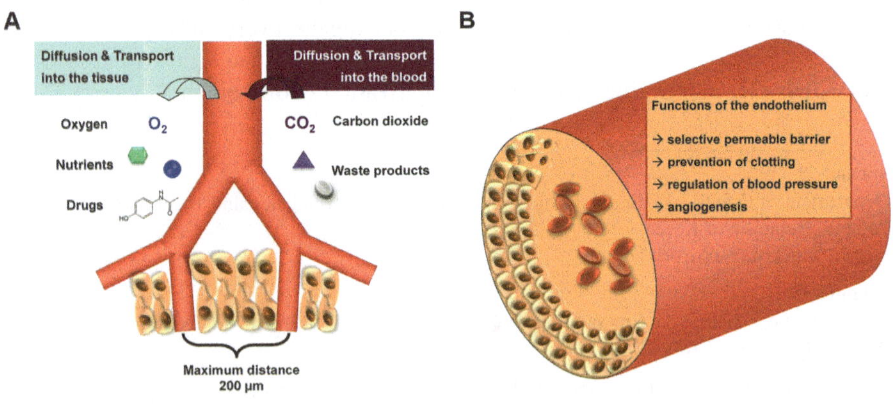

Fig. 4.1 (**a**) Schematic description of diffusion and transport processes in vascularized tissues in vivo. The surrounding tissue is supplied with oxygen, nutrients, and drugs via the vasculature. Waste products and CO_2 are cart away from the tissue into the blood vessels. (**b**) The endothelium inside the vessels accomplishes numerous functions: selective permeable barrier, prevention of clotting, regulation of blood pressure, and angiogenesis (Reproduced with permission from [10])

pertaining to multiple surgical procedures and limited availability or unavailability, especially for elderly patients and patients with cardiovascular disease conditions. As an alternative approach, artificial conduits built from biologically inert and synthetic materials such as expanded polytetrafluoroethylene (ePTFE), polyurethane, and polyethylene terephthalate (PET) are commonly used for large-diameter vessel applications [7, 8]. However, they suffer from low patency rates in small-diameter (<6 mm) applications, aneurysms or atherosclerosis, thrombosis, and recurrent stenosis, especially at anastomotic sites [8, 9].

Additional common therapeutic approaches to induce vascularization or to revasculate the ischemic tissues include a complex orchestra of angiogenic growth factors (e.g., basic fibroblast growth factor (bFGF), vascular endothelial growth factor (VEGF), transforming growth factor beta (TGFβ), hepatocyte growth factor (HGF), platelet-derived growth factor (PDGF), angiopoietin, and VEGF mimetic synthetic peptides) and cell-based therapy using differentiated ECs (e.g., human umbilical vein endothelial cells (HUVECs) or endothelial progenitor cells (EPCs)) [10, 11]. One report suggested that the use of human ECs in cell-based therapy could promote the formation of long-lasting and functional blood vessels in vivo upon implantation into immunodeficient animals [12]. However, injecting stem cells directly in the ischemic area resulted into approximately 90% cell loss, and poor cell viability [13]. Similarly, the delivery of proangiogenic growth factors based upon a bolus injection resulted into a burst release and short period of biological activity within the ischemic zone [14].

Tissue engineering holds promise as an alternative strategy for creating functional tissue constructs to restore the structural and functional properties of damaged or diseased tissues/organs. Tissue-engineered in vitro organ models have received significant attention for tissue regeneration and as in vitro disease models and promising drug testing platforms in recent years [15]. When a tissue-engineered construct is implanted into the body, an inflammatory wound-healing response occurs, causing the release of hypoxia-induced angiogenic growth factor from the cells, thereby triggering spontaneous angiogenesis from the surrounding tissue [16]. Nevertheless, this vascular ingrowth from the native host tissue is only several tenths of micrometers per day. Thus, it is too slow for adequate oxygen transport to the cells in the inner part of the tissue construct, thereby causing cell death and resulting in a failed implant [2]. This is because the cells in the central region of the tissue-engineered construct suffer from hypoxic conditions and start necrosis if they are >200 μm away from a vascular structure [17]. Thus, it is necessary to fabricate tissue-engineered vascularized constructs using advanced techniques in order to (1) reduce the risk of cell necrosis in the construct; (2) promote the maturation of ECs and other cell types in the construct; (3) ensure the efficient diffusion of oxygen, nutrients, and (optionally drugs) and remove the waste products from the cells, thereby preserving the cells' metabolic activity; and (4) accelerate the integration of the engineered construct with the host tissue upon implantation.

In this chapter, we discuss the types of biomaterials and the variants of classical strategies employed so far for vascularizing tissue constructs. We also highlight the advent of the 3D bioprinting technique in developing a vascularized tissue construct with enhanced functionality and list some applications in tissue regeneration and drug testing.

4.2 In Vitro Tissue-Engineered Vascularized Construct

4.2.1 Choice of Biomaterials

In terms of choosing an appropriate biomaterial, factors such as biocompatibility, biodegradability, resistance to thrombosis, elasticity, promotion of angiogenesis, and integration with host blood vessels should be considered [10, 11]. Specifically, the formation of a capillary-like structure, which is crucial for angiogenesis, is dependent on material properties such as stiffness, composition, and microstructure [18] and the presence of cell adhesion motifs [19]. Thus, it is necessary to select a biomaterial that can balance the intricate interplay of these crucial parameters to ensure angiogenesis as well as the fabrication of a 3D construct. The biomaterials for developing an in vitro tissue-engineered vascularized construct can be categorized into naturally derived polymers, synthetic polymers, and the decellularized extracellular matrix (dECM) (i.e., a biomimicry approach).

4.2.1.1 Natural Polymers

The naturally derived polymers are an attractive source because they promote cellular adhesion and infiltration due to the presence of inherent cell adhesion moieties [10, 11]. Additionally, they account for the majority of the proteins present in the extracellular matrix (ECM) of native arterial walls, thereby enhancing the biological interactions between the material and the host [20]. A variety of protein- and polysaccharide-based polymers have been studied for developing vascular grafts and vascularized tissue constructs, including fibrinogen [21], tropoelastin [22], elastin/collagen [23], gelatin [24], chitosan [25], and alginate [26]. Collagen [27], fibrin/fibrinogen [28], and hyaluronic acid [29] have been reported to improve angiogenesis and promote lumen formation. Notably gelatin cannot activate or bind to platelet, which is an important characteristic to be used as a material for vascular tissue engineering [30]. Nevertheless, the solubility and inherently poor mechanical properties of gelatin limit its application in contact with bodily fluids. Similarly, other protein-based polymers, such as collagen and hyaluronic acid, suffer from poor mechanical properties, thereby restricting their use to small-diameter vascular grafts (SDVG) applications as standalone constructs [31]. Traditional crosslinking agents, such as glutaraldehyde and polyepoxides, are commonly used to improve the mechanical properties; however, these are associated with low long-term stability and high cytotoxicity [30]. On the contrary, polysaccharide-based polymers, such as alginate, agarose, and chitosan, although mechanically tougher than protein-based polymers, are devoid of the cell-adhering moieties crucial for cell attachment, proliferation, and motility, and thus, apoptosis may be induced in the encapsulated cells [1, 26]. Hence, synthetic polymers are rapidly gaining attention on these grounds as scaffold materials due to reproducibility in consistent quality, and their properties, such as degradability and elasticity, are easy to adapt.

4.2.1.2 Synthetic Polymers

Synthetic polymers provide more flexibility in terms of molecular design, readily allowing precise control over the critical material properties, such as polymerization, degradability, and mechanical stiffness [5, 11]. Commonly used synthetic polymers for vascular tissue engineering include polycaprolactone (PCL) [32], polylactic acid (PLA)/poly(acrylic acid) (PAA) [33], poly(vinyl alcohol) (PVA) [34], polyethylene glycol (PEG) [35], and poly(L-lactide-*co*-caprolactone) (PLCL) [36]. Polymer such as Pluronic F-127 is commonly used as a fugitive material for creating microchannels in the scaffold to enhance the vascular network [37, 38]. Nevertheless, the use of synthetic polymers is associated with serious concerns, such as incomplete degradation, and they do not innately mimic the native tissue and release of toxic or acidic degradation products. Thus, improving the biocompatibility, material biodegradability, and bioactivity of these synthetic polymers becomes a major point in biomaterials research. To address this challenge, synthetic polymers have been blended with natural polymers to engineer a composite possessing a native tissue-like ECM with controllable biophysical and biochemical properties and tunable mechanical properties. Examples of composite scaffolds include PCL/collagen/elastin [39], poly(glycolic acid) PGA/poly(L-lactic acid) PLLA/collagen [40], and poly(1,8-octanediol-co-citrate) (POC)/collagen/elastin [41]. Nevertheless, the use of this strategy may manifest the disadvantages of both polymers, such as the adverse host tissue inflammatory response of synthetic polymers and the biological variability of natural polymers. Hence, an alternative and more impressive engineering approach is the use of the dECM.

4.2.1.3 dECM: A Biomimetic Approach

The dECM is a naturally derived elastic material, containing growth factors that inherently support cellular functions, such as cell proliferation, migration, and differentiation, and can promote angiogenesis in neighboring tissues [42, 43]. The dECM can either be implanted as an acellular construct or repopulated with cells and cultured before implantation in vivo or used as a dECM hydrogel. For instance, Bader et al. reported repopulating the decellularized porcine vessels with human ECs and myofibroblasts and demonstrated high cell viability in vitro with significantly lower rejection response of the acellular scaffolds in vivo than the untreated tissues [44]. Nevertheless, a major limitation of most of the animal-derived matrices is the lack of reproducibility and standardization. In addition, this strategy is associated with inefficient recellularization at the desired locations and incomplete re-endothelialization of the organ vasculature. Moreover, the decellularization process may cause elastin deformation and degradation, damage the vessel structure with increase in porosity, and decrease in mechanical properties. This may in turn result into vascular graft expansion and even aneurysm formation followed by vascular leakage after recellularization [42, 45]. Therefore, it is necessarily desirable to create a novel class of biomaterials to overcome these shortcomings. dECM-based hydrogels can provide a promising platform for novel cell or biomolecule delivery and regenerative therapy via minimally invasive techniques [46]. Gao et al. reported the use of a novel vascular tissue derived dECM (vdECM)-based bioink for the fabrication of tubular structures encapsulated with EPCs and microcapsules of the proangiogenic drugs (atorvastatin) to treat ischemia [47]. The dECM-based hybrid bioink facilitated the direct fabrication of the tubular construct and provided a favorable microenvironment to promote the proliferation, migration, differentiation, and neovascularization of the cells.

4.2.2 Strategies for Fabricating Tissue-Engineered Vascularized Constructs

The fabrication of vascularized tissue constructs is essential to accelerate rapid and functional anastomosis with the host tissue to ensure the survival of cells in the early post-implantation phase [48]. In earlier studies, two very common and dedicated approaches used to induce vascularization and vessel stabilization included (1) incorporation of angiogenic/vasculogenic growth factors and (2) the use of ECs alone or co-cultured with different types of mural cells into the fabricated scaffold [10]. In this approach, ECs and/or other cell types such as myoblasts or fibroblasts are seeded on the fabricated scaffolds and proangiogenic factors are optionally incorporated followed by culturing them in vitro with the aim of developing a 3D prevascularized tissue construct. Subsequently, the prevascularized construct with the primitive vascular network is implanted into the ischemic site. For instance, a bio-tubular scaffold developed from human dermal fibroblasts and HUVECs co-cultured on an electrospun PCL-based membrane supported collagen remodeling and demonstrated improvement in biomechanical properties up to day 14 [49]. Upon in vivo implantation, the co-cultured electrospun scaffold displayed enhanced infiltration of host cells and formation of thin continuous layers of smooth muscle cells (SMCs). Similarly, proangiogenic growth factors can be incorporated in predesigned biodegradable scaffolds to release the loaded protein in a sustained manner following its degradation [50]. An additional approach to deliver growth factors is the use of transfected cells seeded in the scaffold providing sustained growth factor release and overexpressing the angiogenic factors [51]. Nevertheless, composition and dose of the delivered growth factors are critical parameters for effective angiogenesis [52]. Thus, in order to ensure its controlled delivery and to avoid the possibility of uncontrolled effects at secondary sites, immobilizing factors are being utilized in many studies [53].

In vitro prevascularization is a time saving method allowing the host blood vessels to anastomose with the construct in the marginal area instead of growing into the middle of the engineered tissues. However, they are associated with poor perfusion systems, because the missing in vitro tissue is not micro-surgically connected to the host vasculature [10]. Thus, to obtain efficient diffusion of oxygen and nutrients ensuring the survival of transplanted tissue grafts in vivo, studies have started employing bioreactor systems resulting in enhanced cell differentiation and/or ECM deposition. Dynamic cultures in vitro have been reported to assist in vessel orientation, which is important for the vascularization of aligned tissues, such as muscle, ligaments, and nervous tissues [54]. Advancement in the development of vessel conduits was reported by Syedain et al. who demonstrated the fabrication of an arteriovenous (AV) graft from human neonatal dermal fibroblasts entrapped in bovine fibrin gel, cultured in a pulsed flow bioreactor for 5 weeks, and decellularized before being implanted in a baboon [55]. Under in vitro culture conditions, as the fibroblasts grew, they digested the fibrin and deposited their own ECM composed mainly of collagen. The grafts demonstrated extensive recellularization showing smoothelin-positive SMCs and developed an endothelium on the luminal surface, with no observed calcifications, loss of burst strength, or outflow stenosis. Recently, Landau et al. reported the application of oscillatory strain for 21 days as an easy method for enhancing vessel morphogenesis and stabilization within collagen constructs co-cultured with ECs and fibroblasts [54]. The stimulated vessels within the engineered construct were more clearly developed, aligned, and stabilized. Moreover, following their implantation into athymic nude mice, aligned infiltration of the host vasculature was observed.

In contrast to in vitro prevascularization with the pre-seeded scaffold, the in vivo prevascularization is based on an implantation of acellular scaffold [10]. During the preliminary implantation of a tissue-engineered construct into the host body, de novo vascularization occurs at the construct [10]. This phenomenon of the infiltration of host blood vessels into an acellular macroporous scaffold was first demonstrated back in 1972. Since then, many studies demonstrated the development of perfusable vascular networks in implanted acellular scaffolds. After successful vascularization into the implanted acellular scaffold, it is explanted again and inserted into the ischemic target site [10]. Nevertheless, this approach is associated with at least three surgeries: (1) implantation of the acellular construct, (2) explantation of the vascularized construct, and (3) proximate implantation or insertion of the prevascularized construct at the ischemic site. Another polysurgery approach combining thermo-responsive polymers with ingrowing cells is the fabrication of tissue-engineered cell sheets [56]. The confluent monolayers of cells harvested from thermo-responsive substrate as cell sheets are transplanted into the ischemic target site. Similarly, use of AV loop is an additional prevascularization strategy based on in vivo anastomosis to generate the vascularized tissue [57]. Briefly, a shunt loop between an artery and a vein is formed using a synthetic graft followed by enclosing the complete intrinsic model within a chamber that is either empty or employs a scaffold to be vascularized. The chamber is then placed at a site of rich vascularization in vivo, and after the formation of an arterio-capillary-venule network by the near-by tissues, the prevascularized construct can be transferred to the reconstructive defect. Although, it represents a promising advancement toward the clinical use of a prevascularized construct, the need for several surgical interventions is a serious concern.

While different strategies have been employed to fabricate vascularized 3D scaffolds, there is an urgent demand to develop advanced and more sophisticated fabrication strategy. It should be noted that effective vascularization is inherently associated to intelligent scaffold design. Therefore, smart 3D designs and high-resolution manufacturing technologies are adopted to achieve the tailor-made control of oxygen gradients, nutrient flow regimes, as well as cellular alignment and angiogenesis. Biofabrication combining the 3D printing approach using natural polymers as hydrogels and living cells, termed bioprinting, can serve as a reasonable alternative technique for fabricating tissue-engineered vascularized constructs [58]. The concept of bioprinting can be considered as an extension of rapid prototyping technique to develop anatomically relevant complex tissue constructs in a layer-by-layer manner according to a predefined 3D computer-aided design (CAD) model. It enables the fabrication of realistic 3D tissue/organ model ranging from mm- to cm-sized scale, incorporating different types of cells, biomolecules, and biomaterials simultaneously, to better emulate the native tissue architecture and promote long-term cell viability [58].

4.3 3D Bioprinting of Vascularized Constructs

3D bioprinting aims at the precise deposition of a cell-laden biomaterial into anatomically relevant shaped, complex 3D structures by keeping the cells viable and guiding tissue formation into a desired shape and direction [58]. Based on the working principle, 3D bioprinting can principally be categorized into (1) inkjet-based, (2) laser-assisted, and (3) extrusion-based bioprinting.

4.3.1 Inkjet-Based Bioprinting

Inkjet-based bioprinting is a non-contact, fast-fabrication printing technique that involves depositing millimeter-sized droplet modules of bioink from a nozzle onto a hydrogel substrate or culture dish, thereby allowing them to self-assemble into a structure with interstitial space that is similar to the vascular structure required for the perfusion of medium or blood [59, 60]. In 2005, Kesari et al. from the Boland Laboratory in Texas were the first to bioprint SMCs encapsulated in tubular vascular grafts by inkjet printing a $CaCl_2$ solution into an alginate bath [61]. Following this, Nakamura et al. applied a reverse process of ejecting alginate droplets into a solution of $CaCl_2$ to develop micro-gel droplets as the building blocks that subsequently fabricated an alginate-based tubular structure with 200-μm channels [62]. Later, Pataky et al. reported printing alginate-based bioink into a vascularized structure with different geometries and a bifurcating vein with a diameter of 200 μm showing great biocompatibility with fibroblasts [63]. Similarly, fibrin microstructures encapsulated with human microvascular endothelial cells (HMVECs) were micropatterned by inkjet printing droplets of thrombin solution containing cells into a bath of fibrinogen [59]. The printed HMVECs in conjunction with the fibrin were found to proliferate within 7 days, and after 21 days, they aligned themselves inside the channels of the printed fibrin gel structure forming a confluent lining. Chang et al. bioprinted patterns of microvessel fragments within a collagen gel [64]. However, the mature vascular network did not maintain the pre-aligned orientation of the patterned fragments. Despite the advantages such as low cost, rapid deposition, and precise control over deposition of cells, growth factors, and therapeutic reagents, inkjet-based bioprinting has been limited to only thin constructs (i.e., a few layers) due to instability of the printed layers [65]. Adequate supporting materials and structures are required for free-form fabrication process, as the stacking of droplets or strands is currently difficult to control, thereby limiting the resolution. Additionally, it is associated with challenges pertaining to nozzle clogging because of the small nozzle diameter and thermal and mechanical stress on cells during droplet extrusion, thereby dramatically decreasing the cell viability [66].

4.3.2 Laser-Assisted Bioprinting

Toward the aim of fabricating high-resolution constructs, laser-assisted bioprinting (LAB) is a precise fabrication method enabling the deposition of individual cells to create various constructs [1, 60]. Several other researchers used LAB to pattern HUVECs into predesigned vascular structures [67, 68]. The deposited HUVECs started connecting with each other within 1 day; yet HUVECs alone could not form a stable vascular network. Rather, the deposition of an additional supporting cell layer (i.e., human umbilical vein smooth muscle cells (HUVSMCs)) on top of HUVECs was required to support the structural integrity of the vasculature. In addition, the co-culture of HUVECs and SMCs showed a lumen-like vascular structure with multiple junctions, indicating a synergistic effect on developing their cellular morphologies. The symbiotic interaction between HUVECs and SMCs was crucial for the generation of stable and functional artificial blood vessels with tissue substitutes, as the SMCs limited the overgrowth and migration of ECs through the cell–cell junctions. In another study by Gaebel et al., the laser-induced-forward-transfer (LIFT) cell printing technique was applied to the co-cultured HUVECs and human mesenchymal stem cells (hMSCs) in a defined pattern to fabricate functional polyester urethane urea (PEUU) patches for cardiac regeneration [69]. Patches were cultivated in vitro followed by transplanting in vivo to the targeted zone of infarcted rat hearts. The co-culture of HUVECs and hMSCs promoted enhanced vessel formation with significant functional improvement of the infarcted hearts. Even though LAB enables the uniquely precise deposition of individual cells on the substrate, the ribbon preparation makes the method time-consuming [5]. Moreover, the UV light can damage the cellular structures (i.e., cell membrane, DNA, and proteins), causing cell degeneration and death. In addition, regardless of the light source, the high-energy radicals starting the photopolymerization can react with cellular macromolecules and cause oxidative damage to encapsulated cells, thereby altering the secondary and tertiary structures of cellular proteins [70]. However, this damage can be reduced with the proper choice of crosslinking material.

4.3.3 Extrusion-Based Bioprinting

Extrusion-based bioprinting, an advanced and sophisticated additive manufacturing technique, has been derived from the inkjet-based and LAB techniques into layer-by-layer process that enable the fabrication of 3D structures defined by their CAD model [5]. Current approaches for bioprinting vascularized tissues can be separated into two major categories: direct bioprinting and indirect bioprinting [2]. In 2009, Li et al. demonstrated one of the first attempts to bioprint vascularized tissue constructs by extruding multi-material, cell-laden hydrogel structures for liver applications [71]. Two different cell/polymer suspensions were separately extruded into a low-temperature chamber. Hepatocyte-laden gelatin/alginate/chitosan hydrogel formed the matrix of the tissue construct, and gelatin/alginate/fibrinogen hydrogel mixed with adipose-derived stromal cells lined the vascular channels through the matrix. Thrombin was used to stabilize the multi-material construct, and the surface was crosslinked with glutaraldehyde. The stromal cells in the vascular channels differentiated into ECs with the help of endothelial growth factors, and the hepatocytes maintained their normal metabolism within a 2-week in vitro cell culture study. A more recently reported study showed the fabrication of a 3D cell-printed prevascularized cardiac patch employing pig heart-derived dECM (hdECM) encapsulating multiple cell types including human cardiac progenitor cells (CPCs) and MSCs [72]. Upon in vivo implantation in a rat myocardial infarction (MI) model, cellular infiltration into the area of infarction and enhanced cardiac function were evident with improved vascularization as well as reduced cardiac hypertrophy and fibrosis. However, direct bioprinting using a single axial nozzle is not ideal for creating vascularized tissues because of the lack of multidimensional resolution and the accuracy required for fabricating microvessels [2]. Recently, coaxial or multiaxial printing nozzles were used in direct bioprinting for the preparation of a vascularized structure with microvasculature [2, 5]. This system contained one or multiple channels within the same nozzle tip. Jia et al. demonstrated a novel approach capable of fabricating multilayer and highly organized perfusable vascular constructs using a 3D bioprinter equipped with a multiaxial nozzle system and a formula of bioink containing 4-arm poly(ethylene glycol)-tetra-acrylate (PEGTA), sodium alginate, and gelatin methacryloyl (GelMA) [73]. Two-step crosslinking was used for the preparation of the vascular constructs. The alginate component was ionically crosslinked using a $CaCl_2$ solution that was deposited from both an inner and outer nozzle to maintain the shape of initially bioprinted perfusable tubes. After printing was complete, the constructs were stabilized by subsequent photocrosslinking of both GelMA and PEGTA components; however, alginate was dissolved to increase cell spreading and proliferation. The vascularized constructs showed beneficial biological properties including supporting the attachment and proliferation of encapsulated ECs and stem cells and forming of well-organized and perfusable artificial blood vessels. Recently, Gao et al. reported developing a tubular-like structure using a coaxial bioprinting technique and hybrid bioink composed of 3% vdECM extracted from porcine aortic tissue and 2% (w/v) sodium alginate mixed with EPCs and supplemented with microcapsules of the atorvastatin drug [47]. The results demonstrated enhanced survival and differentiation of stem cells and an increased rate of neovascularization, highlighting its novelty for the treatment of ischemic diseases. The use of a coaxial or multiaxial nozzle provides an efficient way to directly prepare a vascularized structure with a small diameter and high resolution close to 100 mm, which is a step toward overcoming the challenges of manufacturing vascularized tissues through bioprinting. Overall, direct bioprinting is a simple and fast approach to fabricate micron-range capillary networks. However, the disadvantages of direct bioprinting, including the limitation of available bioinks, the inability to create branched vascular structures over a range of scales, and the decrease of cell viability because of the extrusion pressure, are hindering the development of this approach to clinical translation [2]. Another approach to fabricate vascularized tissue constructs is through indirect bioprinting, where the vascular structure is not directly printed by the nozzle but by the removal of a fugitive material [2].

Bioinks, such as gelatin [74], agarose [75], and carbohydrate glass [76], have been used as the fugitive material for fabricating vascularized tissues through indirect bioprinting. For instance, Miller et al. formulated a sacrificial material based on carbohydrates, which was printed in filaments into the fibroblast-laden hydrogel using a syringe and a RepRap Mendel 3D printer followed by the dissolution of carbohydrate filaments to form hollow channels or vascular networks [76]. Seeded HUVECs lined the channels within 9 days of in vitro culturing and formed clear vascular lumens surrounded by the encapsulated fibroblasts, thereby enhancing the cell viability of hepatocytes. Similarly, Bertassoni et al. extruded agarose fibers and cast them within a GelMA hydrogel [75]. The dissolution of the fibers resulted in perfusable microchannels, where the seeded HUVECs formed a confluent lining within 7 days of in vitro culturing. Although these methods provide an easy approach to fabricate prevascularized tissue constructs, the carbohydrate glass filaments display a short lifespan and the physical properties of agarose gel may hamper the printing resolution [60, 77].

Kolesky et al. reported an approach to fabricate tissue-engineered vascularized construct with multiple cell lines and a vascular structure [37, 38]. Pluronic F-127 was used as the fugitive bioink for making the microchannels as the vascular structure. In one approach, Kolesky et al. printed vascularized tissue constructs using four separate bioinks, including fugitive Pluronic F-127, poly(dimethyl siloxane), and two different fibroblast-laden GelMA inks [38]. The printed multi-material construct was further embedded in an acellular GelMA hydrogel, and the fugitive ink was sacrificed to result in empty channels. Seeded HUVECs formed an endothelium on the channels after 9 days of in vitro culturing. The results showed the ability of this method to fabricate both intermediate microvessels and macrovasculature with diameters ranging from 100 μm to 1 mm. In addition, lumen formation was observed during the cell culture, and the built-in vascular network significantly enhanced the cell viability of human neonatal dermal fibroblasts and 10T1/2 fibroblast cells. In another approach as a follow-up study, Kolesky et al. reported the fabrication of 3D cell-laden and vascularized tissues with a thickness >1 cm. Pluronic F-127 and thrombin were mixed as the first bioink for printing the vascular structure [37]. Then, hMSCs, human neonatal dermal fibroblasts, fibrinogen, and gelatin were mixed as the second bioink for printing the bone tissue. A solution composed of fibrinogen, gelatin, $CaCl_2$, and transglutaminase was used as the ECM to cast the whole tissue-engineered construct followed by rapid crosslinking. Pluronic F-127 was evacuated for creating the microchannels as the built-in vascular structure. The HUVECs seeded on the microchannels formed a continuous endothelium, and the encapsulated stem cells and fibroblasts migrated toward the channels to support the vascular network. The final tissue-engineered constructs were further perfused for a period of >6 weeks. Lumen formation was clearly observed after 45 days of perfusion. In addition, the differentiation from hMSCs to osteoblasts and observation of new bone formation were found by the expression of osteocalcin, collagen I, calcium phosphate, and alkaline phosphatase. The advantages of indirect bioprinting include the ability to create branched vascular structures, high printing resolution without concerns of decreased cell viability because of the extrusion pressure, and a wide range of bioinks and fugitive materials [2]. However, removing fugitive material on a large scale and forming a complex structure can be difficult and time-consuming, which is the major drawback to this approach [2]. Additional disadvantages include the requirement for post processing, such as the perfusion of the vascular structure with endothelium cell suspension, whereby consistent cell seeding and endothelialization are not guaranteed [2]. Moreover, Pluronic F-127 is associated with drawbacks such as low strength, slow gelation, weak stability, and rapid degradation, thereby limiting its application in tissue engineering [78]. Additionally, it is complex and requires large amounts of bioprinting material. Hence, efforts are still focused on fabricating 3D and more realistic in vitro vascularized construct that can better emulate the in vivo tissue architecture and cellular behavior.

4.4 Application of Bioprinted Vascularized Tissue Constructs

Bioprinted vascularized 3D tissue constructs have received significant attention due to their promising applications in tissue engineering and drug discovery/testing platforms. While some studies reported the proof-of-concept principles, a few advanced studies demonstrated the potential of 3D bioprinted vascularized constructs in tissue regeneration and as drug testing models.

4.4.1 Tissue Regeneration

Bioprinted vascularized cardiac constructs demonstrated improved therapeutic efficacy by rapid vascularization post-transplantation. A functional 3D-bioprinted PEUU patch consisting of a co-culture of HUVECs and hMSCs and fabricated using the LIFT cell printing technique demonstrated significantly enhanced capillary density and integration into the host myocardium post transplantation into the infarcted zone of rat hearts [69] (Fig. 4.2). When co-cultured with stem cells, HUVECs could increase cell viability and enhance vascular differentiation and wound-healing potential through paracrine effects and membrane interactions, thereby improving angiogenesis. On the other hand, stem cells secrete large amounts of pro-survival factors and ECM, thereby protecting the scaffold-seeded HUVECs and increasing the survival rate of the nascent cardiac tissue. Hence, the dual-cell-printed PEUU cardiac patch not only enabled the cellular migration into the damaged heart but also induced angiogenesis, which resulted in increased myocardial blood flow, thereby increasing cardiac function. Similarly, human CPCs and MSCs mixed with porcine hdECM and growth factors were subjected to extrusion bioprinting to fabricate a multicellular and multilayered prevascularized cardiac patch

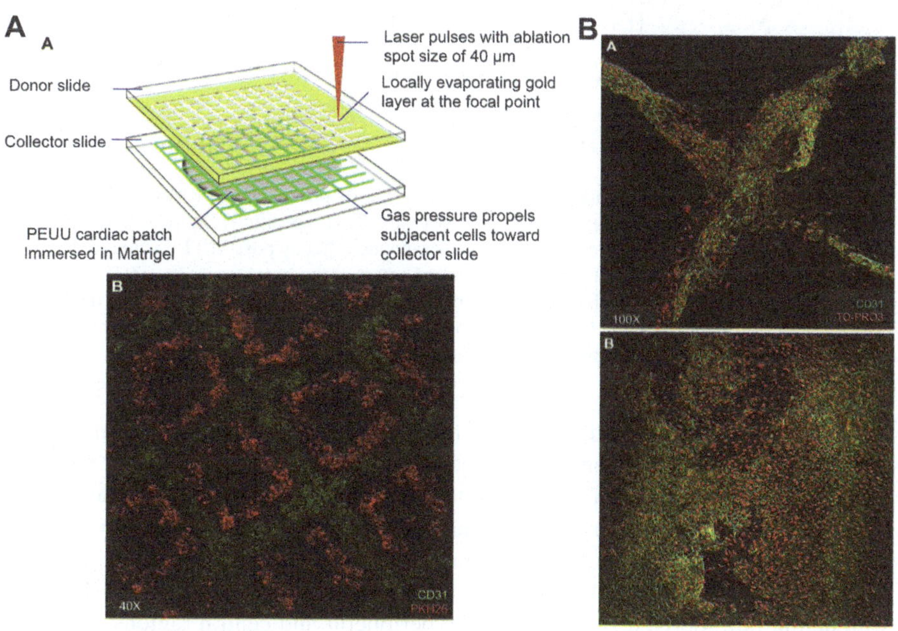

Fig. 4.2 (**a**) Experimental design: *A*. Schematical bioprinting setup based on LIFT. *B*. Arrangement of transferred cells by LIFT observed after 24 h: Human MSC were prestained with PKH26 and patches were stained with polyclonal goat anti-Pecam1 24 h after LIFT to separate grid patterned HUVEC. (**b**) In vitro acceleration of tube forming by LIFT: Representative immunofluorescent micrographs of Pecam1 stained patches treated with co-culture of HUVEC/hMSC 8 days after cell seeding. *A*. Vessel formation in the simulated printed pattern. *B*. Proliferation of cells but marginal vessel formation on the randomly seeded control matrix (Reproduced with permission from [69])

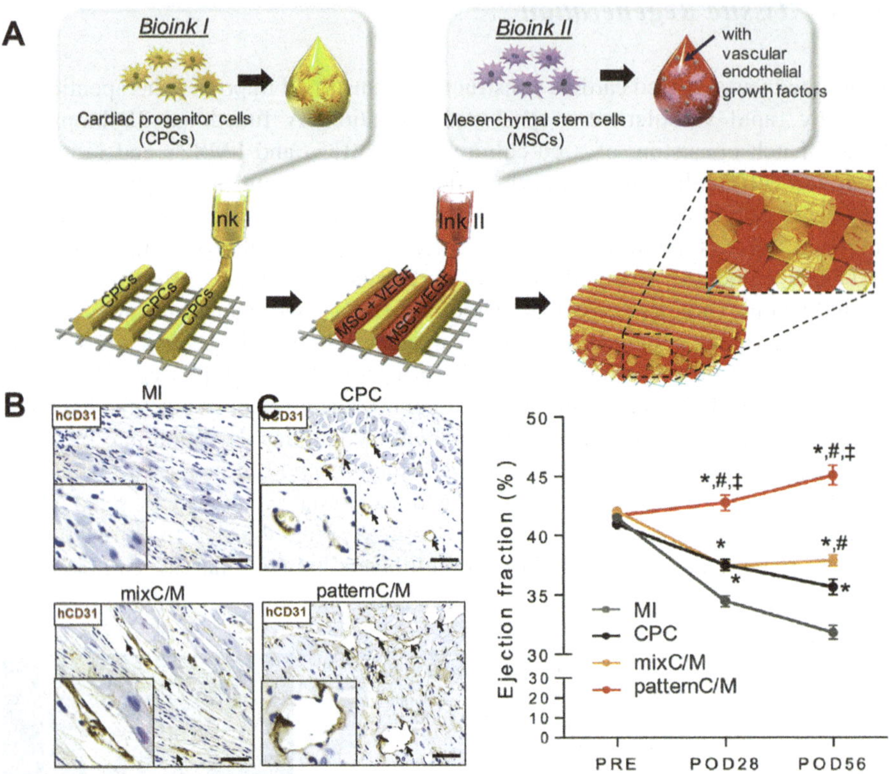

Fig. 4.3 (**a**) Illustrative representation of 3D cell printing of prevascularized stem cell patch. (**b**) Immunohistochemistry results against human-specific CD31 specificity at infarct regions. (**c**) Effects of prevascularized stem cell patch on the therapeutic efficacy post-MI; EF values at baseline and after 4 and 8 weeks. Error bars represent standard errors of the mean (sem) ($^{*}p < 0.05$ compared with MI; $^{\#}p < 0.05$ compared with CPC; $^{\ddagger}p < 0.05$ compared with mix C/M). *MI* Myocardial infarction, *EF* Ejection fraction, *CPC* Cardiac progenitor cells, *C/M* Both CPC and mesenchymal stem cells (MSCs), *POD* Post operation day (Reproduction with permission from [72])

[72] (Fig. 4.3). Upon in vivo implantation of the fabricated cardiac patch in a rat MI model, enhanced cardiac function and cellular infiltration into the infarct area was evident. Additionally, it also demonstrated improved vascularization, reduced cardiac hypertrophy, and fibrosis as well as tissue formation. In another study, Park et al. reported a novel approach using an extrusion-based 3D printing technique to load growth factors (i.e., bone morphogenetic protein-2 (BMP-2) and VEGF) along with dental pulp-derived stem cells in the peripheral and central regions of porous multi-material bioprinted scaffolds, respectively, to induce vascularized bone regeneration [79] (Fig. 4.4). The loaded VEGF stimulated the formation of microvessels throughout the cell-laden scaffolds, and the sequentially released BMP-2 promoted bone regeneration in vivo. These studies highlight the importance of 3D bioprinting in generating new capacities for designing vasculature and patterning proangiogenic factors to accelerate tissue regeneration.

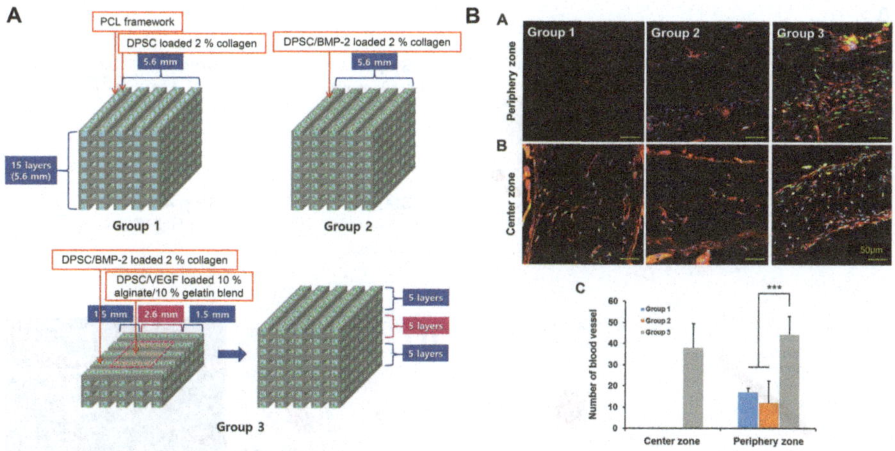

Fig. 4.4 (**a**) Structure designs. Group 1: DPSCs printed structure using 2% collagen. Group 2: DPSC/BMP-2 printed structure using 2% collagen. Group 3: DPSC/dual growth factors printed structure using 2% collagen and 10% alginate/10% gelatin blend. (**b**) Vessel formation in the implanted structures. Observation of positive staining with BSI-Lectin (red), VE-cadherin (green), and DAPI (blue) at the (*A*) center and (*B*) periphery zone of the printed structure of groups 1, 2, and 3. (**c**) Average number of blood vessels per 0.1 mm^2 at three different sites in the implanted structures. Bars: ±1 s.d., $n = 3$; $^{***}p < 0.001$. (Reproduced with permission from [79])

4.4.2 Drug Discovery and Toxicological Screening

3D-bioprinted constructs have also been subjected to extensive research for drug testing and discovery platforms. Massa et al. developed a 3D vascularized liver tissue model to mimic physiological drug diffusion and toxicity testing [77] (Fig. 4.5). Sacrificial bioprinting strategy was employed to successfully develop hollow microchannels in the 3D liver tissue model consisting of HepG2/C3A cells encapsulated in gelatin methacryloyl hydrogel. After seeding HUVECs into the microchannel, the authors reported obtaining a vascularized tissue construct lined by a monolayer of ECs into which nutrients, oxygen, media, and drugs could be introduced easily. Following to it, acetaminophen (APAP) was perfused through the endothelialized channel for 48 h to observe the effects of the endothelial layer on drug administration mechanism. The inclusion of the HUVEC layer created a barrier that delayed the permeability of the drug from the microchannels to the surrounding hydrogel scaffold, thereby protecting the HepG2/C3A cells from toxicity caused by the drug treatment. The study highlighted the pivotal role of an engineered endothelial layer in the drug administration process. The fabricated 3D vascularized liver model is anticipated to be helpful in observing and predicting drug toxicity mechanisms at a microcirculatory level. In another study, a bio-blood-vessel was fabricated using a novel hybrid bioink (mixture of vdECM and alginate) along with a versatile coaxial extrusion-based 3D cell printing technique to effectively deliver EPCs and microcapsules of the proangiogenic drug atorvastatin to the ischemic area [47] (Fig. 4.6).

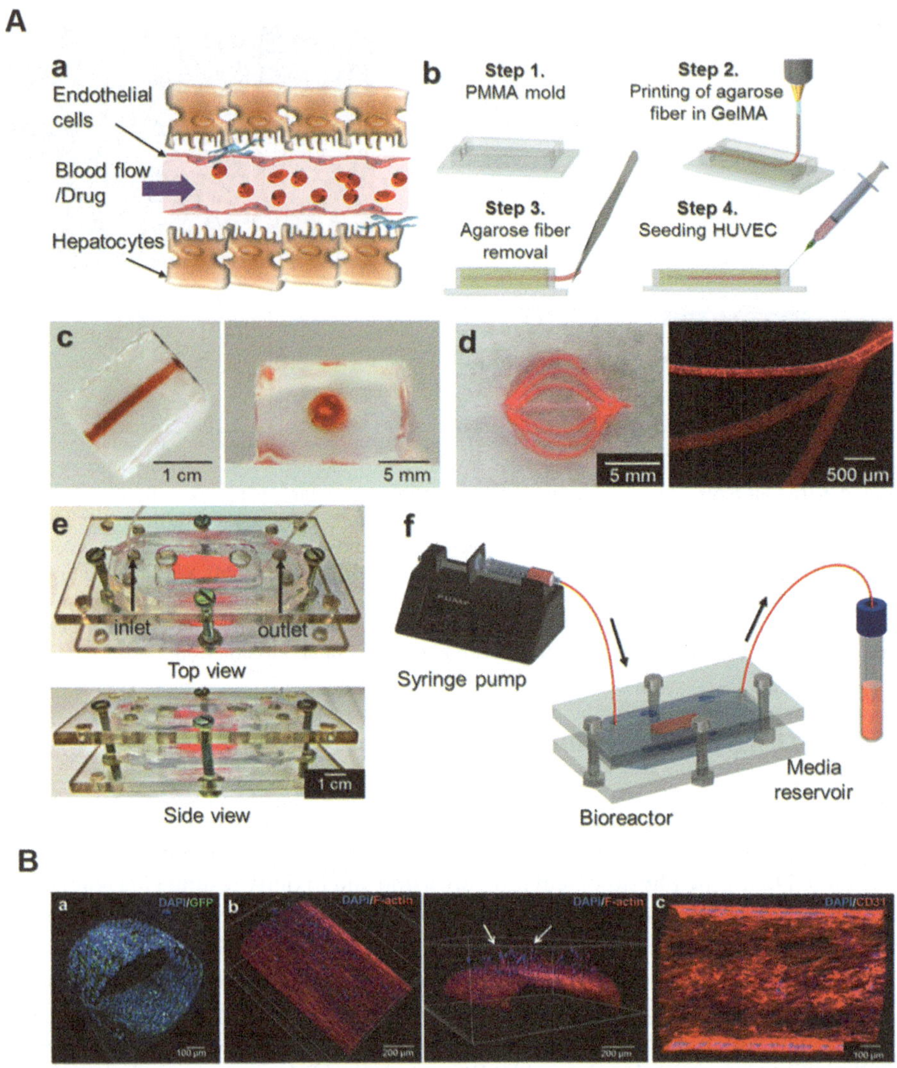

Fig. 4.5 (**a**) (*a*) Schematic of vascularized liver structure. (*b*) Fabrication of a 3D engineered liver tissue construct with a perfusable blood vessel structure. (*c*) Photographic images of the fabricated hydrogel construct with a central hollow microchannel structure and (*d*) curved microchannels embedded inside a circular GelMA structure. Red dye solution was perfused into the hydrogel construct after the molding and crosslinking process. (*e*) Photographs of the resealable bioreactor. (*f*) Illustration of the bioreactor connected to the perfusion module. (**b**) Confocal microscopy images of (*a*) the integrity of the GFP-HUVEC layer in the construct with nuclei staining after 2 days and (*b*) top and cross-sectional views of F-actin/nuclei staining after 7 days which showed endothelial cell invasion and sprouting into the surrounding GelMA hydrogel (arrows). (*c*) Top view of a HUVEC layer immunostained for CD31 (red) and 4′,6-diamidino-2-phenylindole (DAPI) (blue) after 2 days in culture. (Reproduced with permission from [77])

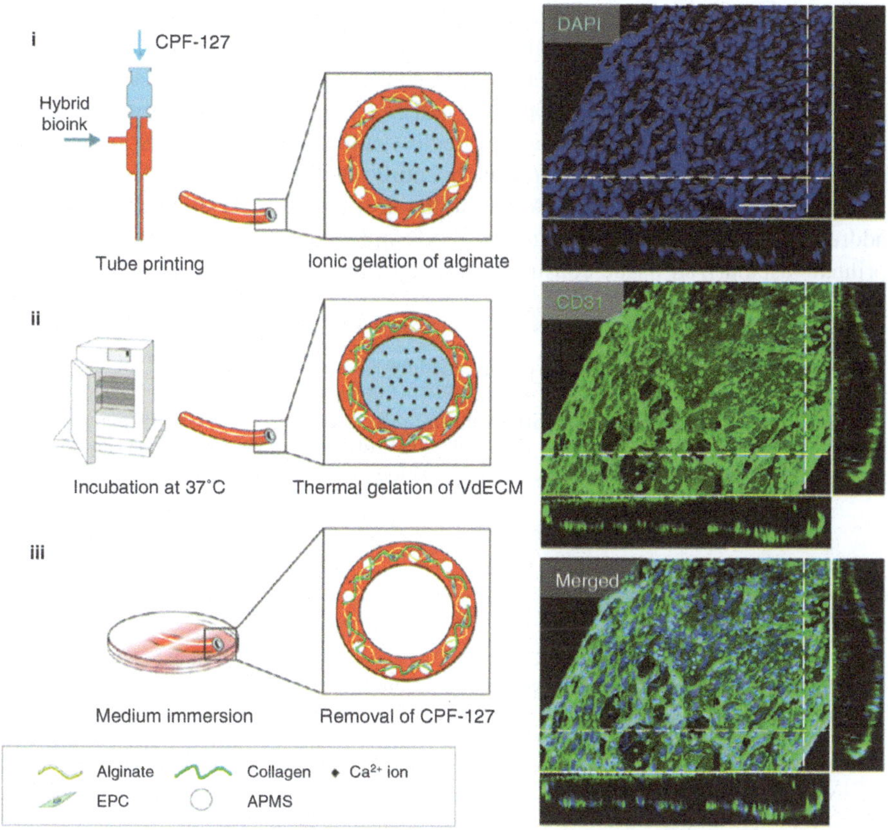

Fig. 4.6 A schematic depiction of the bio-blood-vessel (BBV) fabrication process. The ionic gelation of alginate in the bioink-realized BBV printing. CD31/DAPI staining indicated that the encapsulated endothelial progenitor cells formed a layer of fully differentiated endothelium on the BBV after culturing for 7 days. (Reproduced with permission from [47])

The results demonstrated the enhanced survival and differentiation of stem cells, an increased rate of neovascularization, and remarkable improvement of the ischemic site. This study highlights that the employment of 3D-bioprinted ECM-mediated cell/drug constructs can be a novel therapeutic approach to treat ischemic diseases. Hence, engineered vascularized constructs using advanced bioprinting techniques along with novel bioinks may enable the development of vascular disease models and novel therapeutic approaches toward them.

4.5 Conclusion and Future Perspective

The fabrication of a thick and three-dimensionally arranged vascularized construct is a bottleneck of tissue engineering and regenerative medicine. The requirement in regenerative and therapeutic applications led to the use of the 3D bioprinting technique, thereby opening new avenues for mitigating the challenges posed by the use of conventional strategies. Although the use of bioprinting techniques has demonstrated remarkable progress, numerous challenges have to be met in terms of choosing appropriate bioinks supporting cell viability and functionality and emulating intricate vascular architecture for effective clinical applications. Still, there is a curious paucity of evidence to bioprint an entire functional and complex organ with an efficient vascular system. However, in the near future, after these challenges are addressed, 3D bioprinting technique will be employed to fabricate a functional bioprinted vascularized tissue construct, moving it from the state of proof-of-concept to clinical translation.

Acknowledgements This research was supported by Basic Science Research Program through the National Research Foundation of Korea (NRF) funded by the Ministry of Education (No. 2015R1A6A3A04059015) and the MSIP (Ministry of Science, ICT and Future Planning), Korea, under the "ICT Consilience Creative Program" (IITP-R0346-16-1007) supervised by the IITP (Institute for Information & communications Technology Promotion).

References

1. Hoch E, Tovar GE, Borchers K (2014) Bioprinting of artificial blood vessels: current approaches towards a demanding goal. Eur J Cardiothorac Surg 46(5):767–778
2. Ke D, Murphy SV (2019) Current challenges of bioprinted tissues toward clinical translation. Tissue Eng Part B Rev 25(1):1–13
3. Clarks ER, Clark EL (1939) Microscopic observations on the growth of blood capillaries in the living mammal. Am J Anat 64(2):251–301
4. Kannan RY, Salacinski HJ, Sales K, Butler P, Seifalian AM (2005) The roles of tissue engineering and vascularisation in the development of micro-vascular networks: a review. Biomaterials 26(14):1857–1875
5. Elomaa L, Yang YP (2017) Additive manufacturing of vascular grafts and vascularized tissue constructs. Tissue Eng Part B Rev 23(5):436–450
6. D Levit R (2018) Engineering vessels as good as new? JACC Basic Transl Sci 3(1):119–121
7. Teebken OE, Haverich A (2002) Tissue engineering of small diameter vascular grafts. Eur J Vasc Endovasc 23(6):475–485
8. Wang X, Lin P, Yao Q, Chen C (2007) Development of small-diameter vascular grafts. World J Surg 31(4):682–689
9. Kannan RY, Salacinski HJ, Butler PE, Hamilton G, Seifalian AM (2005) Current status of prosthetic bypass grafts: a review. J Biomed Mater Res B Appl Biomater 74(1):570–581
10. Novosel EC, Kleinhans C, Kluger PJ (2011) Vascularization is the key challenge in tissue engineering. Adv Drug Deliv Rev 63(4–5):300–311
11. Pal A, Vernon BL, Nikkhah M (2018) Therapeutic neovascularization promoted by injectable hydrogels. Bioact Mater 3(4):389–400
12. Schechner JS, Nath AK, Zheng L, Kluger MS, Hughes CC, Sierra-Honigmann MR, Lorber MI, Tellides G, Kashgarian M, Bothwell AL, Pober JS (2000) In vivo formation of complex microvessels lined by human endothelial cells in an immunodeficient mouse. Proc Natl Acad Sci U S A 97(16):9191–9196
13. Li X, Tamama K, Xie X, Guan J (2016) Improving cell engraftment in cardiac stem cell therapy. Stem Cells Int 2016:7168797

14. Tous E, Purcell B, Ifkovits JL, Burdick JA (2011) Injectable acellular hydrogels for cardiac repair. J Cardiovasc Transl Res 4(5):528–542
15. Bhatia SN, Ingber DE (2014) Microfluidic organs-on-chips. Nat Biotechnol 32(8):760–772
16. Laschke MW, Harder Y, Amon M, Martin I, Farhadi J, Ring A, Torio-Padron N, Schramm R, Rucker M, Junker D, Haufel JM, Carvalho C, Heberer M, Germann G, Vollmar B, Menger MD (2006) Angiogenesis in tissue engineering: breathing life into constructed tissue substitutes. Tissue Eng 12(8):2093–2104
17. Jain RK, Au P, Tam J, Duda DG, Fukumura D (2005) Engineering vascularized tissue. Nat Biotechnol 23(7):821–823
18. Kreimendahl F, Kopf M, Thiebes AL, Duarte Campos DF, Blaeser A, Schmitz-Rode T, Apel C, Jockenhoevel S, Fischer H (2017) Three-dimensional printing and angiogenesis: tailored agarose-type I collagen blends comprise three-dimensional printability and angiogenesis potential for tissue-engineered substitutes. Tissue Eng Part C Methods 23(10):604–615
19. Bayless KJ, Salazar R, Davis GE (2000) RGD-dependent vacuolation and lumen formation observed during endothelial cell morphogenesis in three-dimensional fibrin matrices involves the alpha(v)beta(3) and alpha(5)beta(1) integrins. Am J Pathol 156(5):1673–1683
20. Wagenseil JE, Mecham RP (2009) Vascular extracellular matrix and arterial mechanics. Physiol Rev 89(3):957–989
21. Aper T, Wilhelmi M, Gebhardt C, Hoeffler K, Benecke N, Hilfiker A, Haverich A (2016) Novel method for the generation of tissue-engineered vascular grafts based on a highly compacted fibrin matrix. Acta Biomater 29:21–32
22. McKenna KA, Hinds MT, Sarao RC, Wu PC, Maslen CL, Glanville RW, Babcock D, Gregory KW (2012) Mechanical property characterization of electrospun recombinant human tropoelastin for vascular graft biomaterials. Acta Biomater 8(1):225–233
23. Boland ED, Matthews JA, Pawlowski KJ, Simpson DG, Wnek GE, Bowlin GL (2004) Electrospinning collagen and elastin: preliminary vascular tissue engineering. Front Biosci 9:1422–1432
24. Elsayed Y, Lekakou C, Labeed F, Tomlins P (2016) Fabrication and characterisation of biomimetic, electrospun gelatin fibre scaffolds for tunica media-equivalent, tissue engineered vascular grafts. Mater Sci Eng C Mater Biol Appl 61:473–483
25. Kong X, Han B, Wang H, Li H, Xu W, Liu W (2012) Mechanical properties of biodegradable small-diameter chitosan artificial vascular prosthesis. J Biomed Mater Res A 100(8):1938–1945
26. Benning L, Gutzweiler L, Trondle K, Riba J, Zengerle R, Koltay P, Zimmermann S, Stark GB, Finkenzeller G (2018) Assessment of hydrogels for bioprinting of endothelial cells. J Biomed Mater Res A 106(4):935–947
27. Foubert P, Barillas S, Gonzalez AD, Alfonso Z, Zhao S, Hakim I, Meschter C, Tenenhaus M, Fraser JK (2015) Uncultured adipose-derived regenerative cells (ADRCs) seeded in collagen scaffold improves dermal regeneration, enhancing early vascularization and structural organization following thermal burns. Burns 41(7):1504–1516
28. Chung E, Rytlewski JA, Merchant AG, Dhada KS, Lewis EW, Suggs LJ (2015) Fibrin-based 3D matrices induce angiogenic behavior of adipose-derived stem cells. Acta Biomater 17:78–88
29. Tavana S, Azarnia M, Valojerdi MR, Shahverdi A (2016) Hyaluronic acid-based hydrogel scaffold without angiogenic growth factors enhances ovarian tissue function after autotransplantation in rats. Biomed Mater 11(5):055006
30. Ran X, Ye Z, Fu M, Wang Q, Wu H, Lin S, Yin T, Hu T, Wang G (2018) Design, preparation, and performance of a novel bilayer tissue-engineered small-diameter vascular graft. Macromol Biosci 19(3):e1800189
31. Catto V, Farè S, Freddi G, Tanzi MC (2014) Vascular tissue engineering: recent advances in small diameter blood vessel regeneration. ISRN Vasc Med 2014:1–27
32. Fuchs S, Ghanaati S, Orth C, Barbeck M, Kolbe M, Hofmann A, Eblenkamp M, Gomes M, Reis RL, Kirkpatrick CJ (2009) Contribution of outgrowth endothelial cells from human peripheral blood on in vivo vascularization of bone tissue engineered constructs based on starch polycaprolactone scaffolds. Biomaterials 30(4):526–534

33. Gigliobianco G, Chong CK, MacNeil S (2015) Simple surface coating of electrospun poly-L-lactic acid scaffolds to induce angiogenesis. J Biomater Appl 30(1):50–60
34. Yokoyama T, Ohashi K, Kuge H, Kanehiro H, Iwata H, Yamato M, Nakajima Y (2006) In vivo engineering of metabolically active hepatic tissues in a neovascularized subcutaneous cavity. Am J Transplant 6(1):50–59
35. Sultana T, Amirian J, Park C, Lee SJ, Lee BT (2017) Preparation and characterization of polycaprolactone-polyethylene glycol methyl ether and polycaprolactone-chitosan electrospun mats potential for vascular tissue engineering. J Biomater Appl 32(5):648–662
36. Kim SH, Kwon JH, Chung MS, Chung E, Jung Y, Kim SH, Kim YH (2006) Fabrication of a new tubular fibrous PLCL scaffold for vascular tissue engineering. J Biomater Sci Polym Ed 17(12):1359–1374
37. Kolesky DB, Homan KA, Skylar-Scott MA, Lewis JA (2016) Three-dimensional bioprinting of thick vascularized tissues. Proc Natl Acad Sci U S A 113(12):3179–3184
38. Kolesky DB, Truby RL, Gladman AS, Busbee TA, Homan KA, Lewis JA (2014) 3D bioprinting of vascularized, heterogeneous cell-laden tissue constructs. Adv Mater 26(19):3124–3130
39. McClure MJ, Sell SA, Simpson DG, Walpoth BH, Bowlin GL (2010) A three-layered electrospun matrix to mimic native arterial architecture using polycaprolactone, elastin, and collagen: a preliminary study. Acta Biomater 6(7):2422–2433
40. Yokota T, Ichikawa H, Matsumiya G, Kuratani T, Sakaguchi T, Iwai S, Shirakawa Y, Torikai K, Saito A, Uchimura E, Kawaguchi N, Matsuura N, Sawa Y (2008) In situ tissue regeneration using a novel tissue-engineered, small-caliber vascular graft without cell seeding. J Thorac Cardiovasc Surg 136(4):900–907
41. Liu J, Argenta L, Morykwas M, Wagner WD (2014) Properties of single electrospun poly(diol citrate)-collagen-proteoglycan nanofibers for arterial repair and in applications requiring viscoelasticity. J Biomater Appl 28(5):729–738
42. Crapo PM, Gilbert TW, Badylak SF (2011) An overview of tissue and whole organ decellularization processes. Biomaterials 32(12):3233–3243
43. Dew L, English WR, Chong CK, MacNeil S (2016) Investigating neovascularization in rat decellularized intestine: an in vitro platform for studying angiogenesis. Tissue Eng Part A 22(23–24):1317–1326
44. Bader A, Steinhoff G, Strobl K, Schilling T, Brandes G, Mertsching H, Tsikas D, Froelich J, Haverich A (2000) Engineering of human vascular aortic tissue based on a xenogeneic starter matrix. Transplantation 70(1):7–14
45. Zou Y, Zhang Y (2012) Mechanical evaluation of decellularized porcine thoracic aorta. J Surg Res 175(2):359–368
46. Pati F, Jang J, Ha DH, Won Kim S, Rhie JW, Shim JH, Kim DH, Cho DW (2014) Printing three-dimensional tissue analogues with decellularized extracellular matrix bioink. Nat Commun 5:3935
47. Gao G, Lee JH, Jang J, Lee DH, Kong J-S, Kim BS, Choi Y-J, Jang WB, Hong YJ, Kwon S-M, Cho D-W (2017) Tissue engineered bio-blood-vessels constructed using a tissue-specific bioink and 3D coaxial cell printing technique: a novel therapy for ischemic disease. Adv Funct Mater 27(33):1700798
48. Shieh SJ, Vacanti JP (2005) State-of-the-art tissue engineering: from tissue engineering to organ building. Surgery 137(1):1–7
49. Lee B, Shafiq M, Jung Y, Park J-C, Kim SH (2016) Characterization and preparation of bio-tubular scaffolds for fabricating artificial vascular grafts by combining electrospinning and a co-culture system. Macromol Res 24(2):131–142
50. Karal-Yilmaz O, Serhatli M, Baysal K, Baysal BM (2011) Preparation and in vitro characterization of vascular endothelial growth factor (VEGF)-loaded poly(D,L-lactic-co-glycolic acid) microspheres using a double emulsion/solvent evaporation technique. J Microencapsul 28(1):46–54
51. Geiger F, Lorenz H, Xu W, Szalay K, Kasten P, Claes L, Augat P, Richter W (2007) VEGF producing bone marrow stromal cells (BMSC) enhance vascularization and resorption of a natural coral bone substitute. Bone 41(4):516–522

52. Zisch AH, Lutolf MP, Hubbell JA (2003) Biopolymeric delivery matrices for angiogenic growth factors. Cardiovasc Pathol 12(6):295–310
53. Saik JE, Gould DJ, Watkins EM, Dickinson ME, West JL (2011) Covalently immobilized platelet-derived growth factor-BB promotes angiogenesis in biomimetic poly(ethylene glycol) hydrogels. Acta Biomater 7(1):133–143
54. Landau S, Ben-Shaul S, Levenberg S (2018) Oscillatory strain promotes vessel stabilization and alignment through fibroblast YAP-mediated mechanosensitivity. Adv Sci (Weinh) 5(9):1800506
55. Syedain ZH, Graham ML, Dunn TB, O'Brien T, Johnson SL, Schumacher RJ, Tranquillo RT (2017) A completely biological "off-the-shelf" arteriovenous graft that recellularizes in baboons. Sci Transl Med 9(414):eaan4209
56. Sekine H, Shimizu T, Hobo K, Sekiya S, Yang J, Yamato M, Kurosawa H, Kobayashi E, Okano T (2008) Endothelial cell coculture within tissue-engineered cardiomyocyte sheets enhances neovascularization and improves cardiac function of ischemic hearts. Circulation 118(14 Suppl):S145–S152
57. Mian RW, Morrison WA, Hurley JV, Penington AJ, Romeo R, Tanaka Y, Knight KR (2000) Formation of new tissue from an arteriovenous loop in the absence of added extracellular matrix. Tissue Eng 6(6):595–603
58. Chang CC, Boland ED, Williams SK, Hoying JB (2011) Direct-write bioprinting three-dimensional biohybrid systems for future regenerative therapies. J Biomed Mater Res B Appl Biomater 98(1):160–170
59. Cui X, Boland T (2009) Human microvasculature fabrication using thermal inkjet printing technology. Biomaterials 30(31):6221–6227
60. Miri AK, Khalilpour A, Cecen B, Maharjan S, Shin SR, Khademhosseini A (2019) Multiscale bioprinting of vascularized models. Biomaterials 198:204–216
61. Kesari P, Xu T, Boland T (2005) Layer-by-layer printing of cells and its application to tissue engineering. Mater Res Soc Symp P 845:111–117
62. Nakamura M, Nishiyama Y, Henmi C, Iwanaga S, Nakagawa H, Yamaguchi K, Akita K, Mochizuki S, Takiura K (2008) Ink jet three-dimensional digital fabrication for biological tissue manufacturing: analysis of alginate microgel beads produced by ink jet droplets for three dimensional tissue fabrication. J Imaging Sci Technol 52(6):060201
63. Pataky K, Braschler T, Negro A, Renaud P, Lutolf MP, Brugger J (2012) Microdrop printing of hydrogel bioinks into 3D tissue-like geometries. Adv Mater 24(3):391–396
64. Chang CC, Krishnan L, Nunes SS, Church KH, Edgar LT, Boland ED, Weiss JA, Williams SK, Hoying JB (2012) Determinants of microvascular network topologies in implanted neovasculatures. Arterioscler Thromb Vasc Biol 32(1):5–14
65. Boland T, Xu T, Damon B, Cui X (2006) Application of inkjet printing to tissue engineering. Biotechnol J 1(9):910–917
66. Gudapati H, Dey M, Ozbolat I (2016) A comprehensive review on droplet-based bioprinting: past, present and future. Biomaterials 102:20–42
67. Guillotin B, Souquet A, Catros S, Duocastella M, Pippenger B, Bellance S, Bareille R, Remy M, Bordenave L, Amedee J, Guillemot F (2010) Laser assisted bioprinting of engineered tissue with high cell density and microscale organization. Biomaterials 31(28):7250–7256
68. Wu PK, Ringeisen BR (2010) Development of human umbilical vein endothelial cell (HUVEC) and human umbilical vein smooth muscle cell (HUVSMC) branch/stem structures on hydrogel layers via biological laser printing (BioLP). Biofabrication 2(1):014111
69. Gaebel R, Ma N, Liu J, Guan J, Koch L, Klopsch C, Gruene M, Toelk A, Wang W, Mark P, Wang F, Chichkov B, Li W, Steinhoff G (2011) Patterning human stem cells and endothelial cells with laser printing for cardiac regeneration. Biomaterials 32(35):9218–9230
70. McCall JD, Anseth KS (2012) Thiol-ene photopolymerizations provide a facile method to encapsulate proteins and maintain their bioactivity. Biomacromolecules 13(8):2410–2417
71. Li SJ, Xiong Z, Wang XH, Yan YN, Liu HX, Zhang RJ (2009) Direct fabrication of a hybrid cell/hydrogel construct by a double-nozzle assembling technology. J Bioact Compat Polym 24(3):249–265

72. Jang J, Park HJ, Kim SW, Kim H, Park JY, Na SJ, Kim HJ, Park MN, Choi SH, Park SH, Kim SW, Kwon SM, Kim PJ, Cho DW (2017) 3D printed complex tissue construct using stem cell-laden decellularized extracellular matrix bioinks for cardiac repair. Biomaterials 112:264–274
73. Jia W, Gungor-Ozkerim PS, Zhang YS, Yue K, Zhu K, Liu W, Pi Q, Byambaa B, Dokmeci MR, Shin SR, Khademhosseini A (2016) Direct 3D bioprinting of perfusable vascular constructs using a blend bioink. Biomaterials 106:58–68
74. Lee VK, Lanzi AM, Haygan N, Yoo SS, Vincent PA, Dai G (2014) Generation of multi-scale vascular network system within 3D Hydrogel using 3D bio-printing technology. Cell Mol Bioeng 7(3):460–472
75. Bertassoni LE, Cecconi M, Manoharan V, Nikkhah M, Hjortnaes J, Cristino AL, Barabaschi G, Demarchi D, Dokmeci MR, Yang Y, Khademhosseini A (2014) Hydrogel bioprinted microchannel networks for vascularization of tissue engineering constructs. Lab Chip 14(13):2202–2211
76. Miller JS, Stevens KR, Yang MT, Baker BM, Nguyen DH, Cohen DM, Toro E, Chen AA, Galie PA, Yu X, Chaturvedi R, Bhatia SN, Chen CS (2012) Rapid casting of patterned vascular networks for perfusable engineered three-dimensional tissues. Nat Mater 11(9):768–774
77. Massa S, Sakr MA, Seo J, Bandaru P, Arneri A, Bersini S, Zare-Eelanjegh E, Jalilian E, Cha BH, Antona S, Enrico A, Gao Y, Hassan S, Acevedo JP, Dokmeci MR, Zhang YS, Khademhosseini A, Shin SR (2017) Bioprinted 3D vascularized tissue model for drug toxicity analysis. Biomicrofluidics 11(4):044109
78. Wu W, DeConinck A, Lewis JA (2011) Omnidirectional printing of 3D microvascular networks. Adv Mater 23(24):H178–H183
79. Park JY, Shim JH, Choi SA, Jang J, Kim M, Lee SH, Cho DW (2015) 3D printing technology to control BMP-2 and VEGF delivery spatially and temporally to promote large-volume bone regeneration. J Mater Chem B 3(27):5415–5425

Chapter 5
3D Bioprinting in Clinical Cardiovascular Medicine

Alexander Cetnar, Martin Tomov, Andrea Theus, Bryanna Lima, Agastya Vaidya, and Vahid Serpooshan

Abstract 3D bioprinting is a form of additive manufacturing tailored toward creating biological constructs with precise spatial control. As an extension of conventional 3D printing with a variety of materials such as polymers, ceramics, and metals, 3D bioprinting focuses on building viable, biomimetic products that can be used to replicate, improve, or substitute functional tissues. Driven by the field of tissue engineering, advancements in 3D bioprinting have enabled greater print resolutions, more customizable bioinks, and faster biomanufacturing speeds, which are critical when handling delicate biological substances. To date, researchers and engineers have creatively employed 3D bioprinting to combat cardiovascular disease, the most prevalent cause of death in the Western world. In the realm of cardiovascular medicine, 3D bioprinting has seen manifold applications including surgical models, cardiac patches, computational and theoretical models, heart valves, and stents. These technologies vary in terms of their extent of development, ranging

A. Cetnar · B. Lima
Department of Biomedical Engineering, Emory University and Georgia Institute of Technology, Atlanta, GA, USA
e-mail: alexander.cetnar@emory.edu; blima6@gatech.edu

M. Tomov · A. Theus
Department of Biomedical Engineering, Emory University and Georgia Institute of Technology, Atlanta, GA, USA

Department of Pediatrics, Emory University School of Medicine, Atlanta, GA, USA
e-mail: martin.lyubomirov.tomov@emory.edu; andrea.theus@emory.edu

A. Vaidya
Department of Biology, Emory University, Atlanta, GA, USA

Emory University and Georgia Institute of Technology, Atlanta, GA, USA
e-mail: agastya.tushar.vaidya@emory.edu

V. Serpooshan (✉)
Department of Biomedical Engineering, Emory University and Georgia Institute of Technology, Atlanta, GA, USA

Department of Pediatrics, Emory University School of Medicine, Atlanta, GA, USA

Children's Healthcare of Atlanta, Atlanta, GA, USA
e-mail: vahid.serpoosahan@bme.gatech.edu

M. Guvendiren (ed.), *3D Bioprinting in Medicine*,
https://doi.org/10.1007/978-3-030-23906-0_5

from in vitro modeling to clinical therapies. While surgical models are most widely used in a clinical setting, other bioprinted models are rapidly developing with promising results. Overall, this chapter focuses on the clinical applications of 3D bioprinting aimed toward understanding, augmenting, or replacing cardiovascular tissues and organs.

Keywords 3D bioprinting · Additive manufacturing · Cardiovascular tissue engineering · Stents · Cardiac valve · Computational modeling · Surgical model

5.1 Introduction

3D bioprinting primarily aims for biomanufacturing of clinically applicable products that can replace diseased/damaged tissues or organs in vivo or creating biomimetic platforms to model various diseases in vitro. Clinical applications include a variety of direct regenerative approaches (e.g., printed tissue patches) and use as supplemental tools to improve current and future patient care methods (e.g., surgical models) (Fig. 5.1). Recent advances in bioprinting have made it possible to fabricate complex, patient-specific tissue architectures while maintaining the viability and function of multiple cell types that recapitulate the cellular and extracellular niche of the target organ/tissues [1]. While there remain some challenges for the clinical application of bioprinted constructs, development of new organ-specific

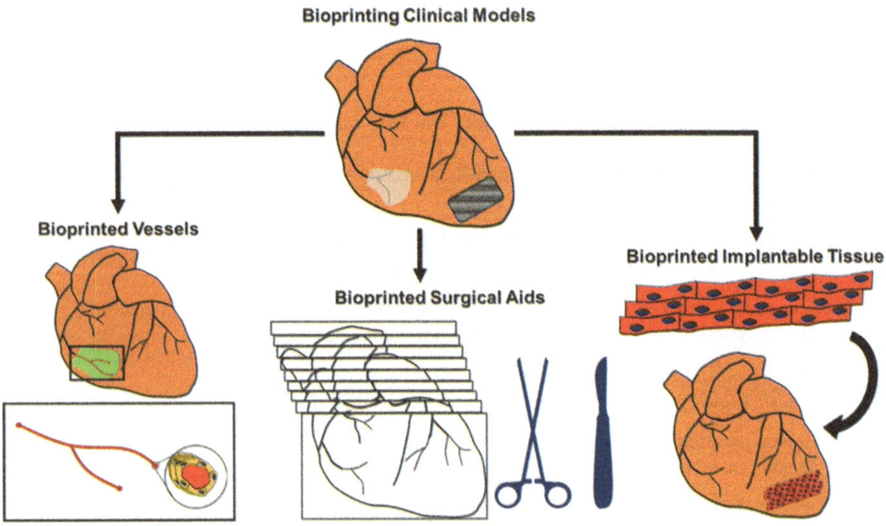

Fig. 5.1 Heart conditions can be addressed via bioprinting that can supplement surgical performance and aid in regenerative medicine. Bioprinted vessels can replace damaged or blocked native vasculature of the heart (left). Improved surgical aids can be developed via a combination of bioprinting and 3D printing approaches (middle). Bioprinted implantable tissue constructs (e.g., cardiac patches) can salvage heart structure and/or performance post infarct or with a congenital condition (right)

bioinks, state-of-the-art medical imaging technologies such as multi-contrast computed tomography (CT), and hybrid 3D bioprinting/printing approaches could be important steps toward translating bioprinting into a clinical setting.

To date, additive manufacturing, and in particular, 3D bioprinting have found rapidly growing applications in the fields of cardiovascular tissue engineering and regenerative medicine [2–5]. By providing a precise spatial control on the cell-biomaterial microenvironment, bioprinting enables recapitulating the complex physiomechanical, chemical, and biological cues of the native heart tissue [2, 6]. While the majority of cardiovascular tissue bioprinting efforts have been focused on restoring the anatomical and structural features of the tissues/organs, new research developments enable bioengineering of functional cardiac constructs [3, 7]. In addition to in vivo regenerative therapies, 3D printed products are increasingly used to enhance diagnosis and treatment of various cardiovascular diseases [8] (Fig. 5.1).

5.2 3D Bioprinted Surgical Models of Cardiac Disease

Bioprinting has grown from a strict research and development tool into a viable approach to generate surgical and clinical models of cardiac disease [6]. The major push that propelled bioprinting is the introduction of reliable, robust bioprinters and functional bioinks, which enable generation of a wide range or practical biomimetic constructs. Specifically, this technology is well suited to produce functional tissue, replacement vasculature for the heart, and high-fidelity anatomical models that can aid in surgical preparation and training [9–11]. Borrowing from the more established 3D printing field, bioprinting can specifically support the production of anatomical models to be used in cardiac surgery, such as surgical guides, templates, and stents (Fig. 5.2) [12]. Further, manufacturing implants such as tissue patches or replicates of the target area for direct organ repair are also possible via bioprinting [13]. Advantages of such 3D bioprinted tools and models include improvement of pre-operative planning, specifically enhancing the accuracy of the used techniques and available practice to perform complex surgeries prior to the operation. This can additionally save time in the operating room, increasing the odds for surgical success [14]. Still, several challenges remain that currently hamper widespread use of bioprinted surgical models and tools in the clinics. The accuracy of generated models is not always sufficient for the purpose, if their desirable characteristics are to be preserved. Depending on the mode of bioprinting, it can be a significant time commitment to generate an accurate biomimetic tool, which may be unfeasible for emergency surgery, though less of an issue when tackling chronic or diagnosed heart conditions. Finally, the relatively high costs associated with the hardware and software (bioprinters and professional CAD programs) and consumables (bioinks, cells, and molecules) is another limitation to routine use of bioprinting as a surgical aid at present [15].

Another application where bioprinting can be successfully translated into a surgical or clinical use is in the design of bioprinted tissue patches. Such printed tissue

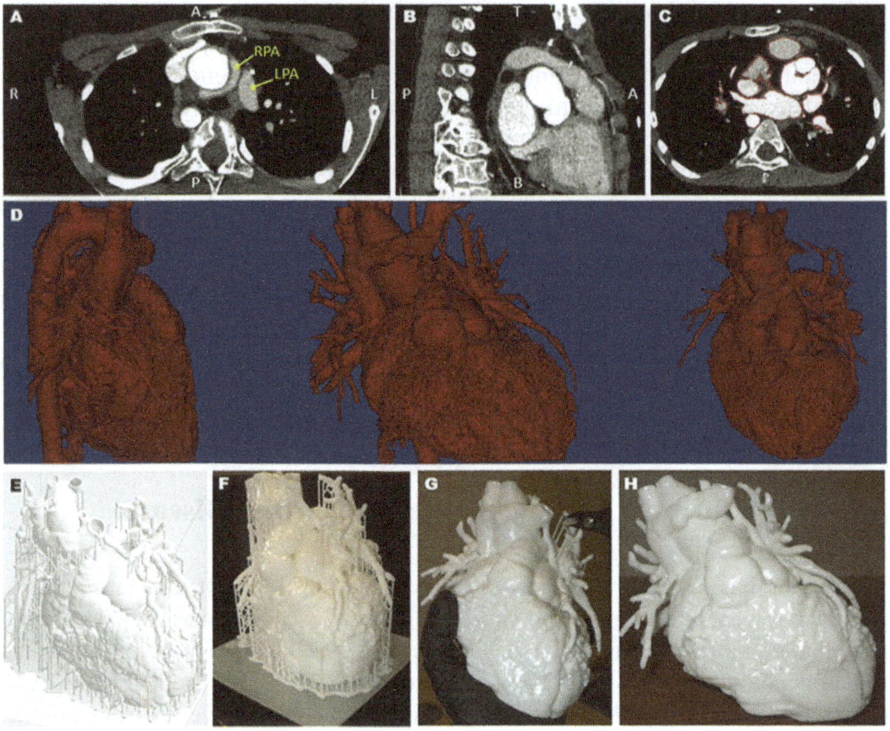

Fig. 5.2 Patient-specific model of heart segmented from computed tomography scans (**a–c**) into a 3D model (**d**). Sacrificial support scaffolds are generated (**e**), printed (**f**), and removed (**g**) to generate the finished heart model (**h**) [16]

models can be used as a surgical training tool, providing a safe, reproducible, and patient-specific platform on which novel devices or techniques could be tested. This could in turn help avoid further complications for patients or relying on an imperfect animal surrogate. Successful cardiovascular tissue models require tight controls over a range of physical parameters that allow to tailor a bioink to a specific clinical or surgical application. Bioink properties (e.g., viscosity, crosslinking mechanism, and resulting stiffness), mass transfer properties (e.g., diffusion and permeability), and functional modifications like biodegradability are some of the parameters that can be tuned to produce a faithful representation of the organ or tissue [17–20].

5.3 3D Bioprinted Cardiac Patches

With over 1.5 million cases of myocardial infarction (MI) each year, there is a demand for patient-specific heart tissue that aims to repair damaged regions [21]. A variety of approaches have been taken to produce cardiac patch devices [22–26].

The integration of human induced-pluripotent stem cell (hiPSC) technology has become a recognized modality for personalized heart tissue engineering [27–30]. Additional methods that are associated with manufacturing implantable cardiac patches are based on cells deriving from mesenchymal stem cells [31], secreted exosomes [32], and decellularized structures [33].

To date, a variety of 3D bioprinting approaches and a growing number of cardiac cell types are being used to manufacture functional cardiac patch systems (Fig. 5.3). A significant number of 3D bioprinted tissues employ cell-based therapeutic processes to improve cardiac function and salvage damaged tissue [34, 35]. Most of these engineered patches recognize the importance of non-muscle cells in myocardial structure and function, such as cardiac fibroblasts (FBs) and vascular cells [36]. Co-culture of cardiomyocytes (CMs) with endothelial cells (ECs), for example, takes strides toward implementing vascularization within constructed tissue architecture—a current focus in cardiac patch bioprinting [5, 37]. Other advantage of using multi-lineage cardiac cells, particularly during in vivo implantation, is the generation of natural tissue components (extracellular matrix or ECM) that are optimal for CM attachment and function [38]. Results from one study that performed in vivo implantation of bioprinted cardiac patches, composed of hiPSC-CMs and small proportions of human adult ventricular FBs and umbilical vein ECs, showed high cell density, but a lack of vascularization on the nude rat hearts [37]. A different experiment in which 3D bioprinted tissues with hiPSC-CMs, FBs, and ECs were tested, showed that maximum vascularization can be achieved by arranging the 3D fibers in a Janus geometry spatial organization [39]. Integration of vascularization in 3D bioprints opens new vistas of opportunities for translational application of cardiac patches with more biomimetic vascular networks [5, 40]. Cardiovascular tissue bioprinting is most commonly performed using naturally derived hydrogels as bioink. Synthetic bioinks are also being explored [41]. Naturally derived hydrogels offer cell viability and function, but are often associated with low resolution, poor handling, and inconsistency between batches. Alternatively, synthetic bioinks offer greater physical integrity and allows for more controlled physiochemical

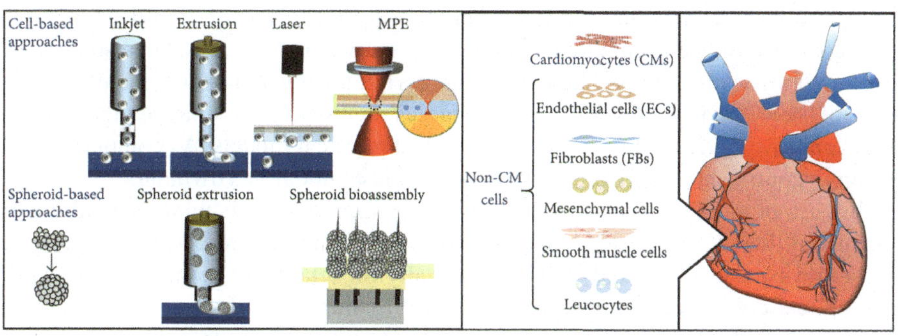

Fig. 5.3 Schematic illustration of various 3D bioprinting approaches used for manufacturing cardiac tissues (left) and specific cell types that reside in cardiac tissue (right) [42]

properties. Synthetic bioinks, however, may provide less accurate biomimicry, and insufficient support for cellular attachment, growth, and function [24].

Several studies have suggested that the distinct electrophysiological properties of nonhuman heart differ greatly from those of the human heart, causing a limitation that is inherent to most types of cardiac patch research [43]. Nevertheless, iPSC-CM-integrated cardiac patches, paired with pieces of decellularized heart ECM, have exhibited beating activity and electrophysiology comparable to those of the human heart muscle [38]. Other attempts to recapitulate human heart tissue in vitro include manufacturing (e.g., bioprinting) relatively larger and thicker human cardiac muscle patches (4 cm × 2 cm × 1.25 mm) [44]. The patch containing hiPSC-CM and hiPSC-smooth muscle cells (SMCs) demonstrated improved cardiac function and decreased infarct wall size and regional wall stress [44].

5.4 3D Bioprinting in Computational and Theoretical Modeling

3D printing provides a viable method for rapidly prototyping biomimetic fluid flow systems for physical flow visualization and computational fluid dynamics (CFD) validation. CFD models of cardiovascular fluids, primarily blood, have improved significantly in terms of their temporal and spatial resolution over the past few decades [45]. Improved imaging modalities, faster computing capabilities, and sleeker solving methods, among others, have enhanced the accuracy and relevance of CFD in clinical settings [46]. However, even the best analytical solvers need to be physically justified to ensure that the predictions made replicate real flow patterns, especially when turbulence is involved [47].

Cardiovascular applications of 3D printing have been studied in an array of diseased and surgically modified geometries. Variations in vessel cross-sectional surface areas in cerebral aneurysms [48], aortic stenosis [49], coarctations [50], and hepatic large vessels [51] have been investigated in normal and abnormal morphologies. Additionally, hemodynamic results of stents [52], embolic coils [53], and cavopulmonary connections [54] on neighboring flows are of significance since they not only study current effects of these corrective interventions, but also provide exploratory in vitro setups. These allow for earlier testing and a more detailed analysis of new devices. Overall, these studies demonstrated relative agreements between flow patterns of CFD simulations and physical models. As computing and imaging continue to improve, errors between these modalities dwindle.

In vitro replication of in vivo hemodynamic scenarios starts with medical imaging modalities such as CT [55], magnetic resonance imaging (MRI) [56], ultrasound and intravascular ultrasound (IVUS) especially for fetal heart monitoring [57, 58], and optical coherence tomography (OCT) [59]. Anatomical features are then segmented to stack 2D images into a 3D construct and meshed through software like 3D Slicer, Vascular Modeling Toolkit, or SimVascular. These files can then be 3D printed or computationally modeled in a CFD software package (e.g., ABAQUS, FLUENT, HARVEY), which typically employs a finite element or lattice Boltzmann solver [45, 60]. Advantageously, the versatility of using the same 3D model to both physically and computationally model a patient-specific anatomical structure allows for consistency during comparison. To visualize fluid flows, many investigators choose to use particle image velocimetry (PIV) in their in vitro models, which captures videos of illuminated (fluorescent) microparticles (beads) and traces their local movement, providing excellent spatial resolution in cross section [54, 61]. Doppler ultrasonography [49] and 4D MRI [51, 62, 63] are also capable of measuring flow velocity, which can provide consistency if both patient and model are imaged with the same modality.

3D printing techniques differ and are typically optimized depending on desired material (e.g., flexibility or optical clarity), fabrication time, cost, resolution, and patient geometry complexity. Specifically, for CFD analyses, inkjet printing is the most common with printed accuracies up to 0.125 mm [48, 49, 52, 64]. Clearing scaffolding for print integrity can be challenging though [48]. Stereolithography (for stiff prints) [54] and laser sintering (for compliant constructs) [51] have provided similar fidelity of 0.1–0.15 mm, respectively. Lost wax molds have been used to generate urethane molds of high optical clarity that, when matched with a liquid of similar refractive index, show no appearance [50, 53]. Deposition extruded positive casts have served to create negative casts for tortuous flow geometries [61]. Both lost wax and deposition modalities can print within 0.2 mm of desired resolution.

While 3D printing has enabled high resolution replicates of clinical flow conduits for in vitro and CFD purposes, limitations exist. Due to the high computational demands of CFD, it is often regulated to specific regions, rather than a full-body vasculature. Not only printing microvasculatures of 1–10 μm, but reliably visualizing them is a challenge too. Since blood vessels are not simply passive vessels, replicating physiological flow mechanics is not trivial when accounting for blood rheology, pulsatility, and tissue compliance and active contractility [51]. Thus far, more traditional 3D printing has been applied for CFD applications, but similar applications of 3D bioprinting have yet to be augmented.

5.5 3D Bioprinted Heart Valves

Heart valve diseases are a serious dilemma worldwide with nearly 80,000 heart valve replacements occurring annually [6, 65]. Current treatment options are valve replacements with mechanical valves [66], bioprosthetic valves from porcine or bovine pericardium [67], or the Ross Procedure [68]. However, there are several downsides to each of these procedures. Mechanical valves often cause problems as a foreign body within the heart. Patients receiving these valves have to be on heavy medication for anticoagulation and immunosuppression for the rest of their lives [24]. Contrarily, bioprosthetic valves deteriorate over a shorter amount of time but can be placed less invasively, reducing patient morbidity [67]. To improve valve longevity and functionality, tissue engineering approaches including decellularization, molded or structured scaffolding, electrospinning, or 3D bioprinting are being explored. Being able to 3D bioprint a biologically compatible heart valve would reduce time, costs, and risks that normally occur with a traditional valve replacement [69].

3D bioprinting allows for accurate replication of heart valves by scanning the 3D conduit, converting it to a STL file for post-processing and printing, and printing a cellular or acellular construct (Fig. 5.4). Bioprinting technology has greatly evolved over the past decade with new approaches allowing for manufacturing heart valves containing more cell types and biomaterials, as well as enabling more patient specificity [70, 71]. The biomimicry of engineered valves is rapidly improving, allowing for greater mechanical integrity, functional longevity, and cell viability [71]. In the future, bioprinting could create heart valves that are able to self-repair or grow as the patient's body grows [72]. The long-term goal for most heart valve tissue engineering studies is to create a valve that would be fully integrated with the host tissue to ensure long-term functionality [14]. Bioprinted valves are also being used in drug screening applications. For example, valve models, bioprinted using encapsulated human valvular interstitial cells (VICs) and exposed to osteogenic media, demonstrated enhanced micro-calcification. Such models can serve as an in vitro platform to study the pathogenesis of calcific aortic valve disease [70].

There are different hydrogels and biomaterials used for 3D printing of heart valves. Some commonly used bioinks include poly-ethylene glycol-diacrylate (PEG-DA), gelatin methacrylate (gelMA), and methacrylated hyaluronic acid (Me-

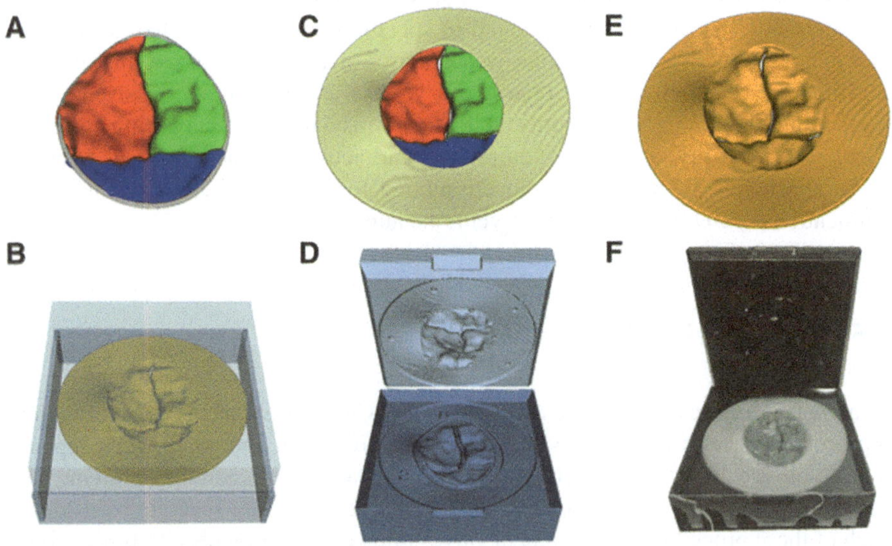

Fig. 5.4 Process of biomanufacturing mitral valve through direct 3D printing: segmentation (**a**), skirt addition (**b**), 3D model (**c**), template mold for 3D printing (**d**), final rendering (**e**), and printed valve within mold [73]

HA) [74, 75]. PEG-DA bioinks supplemented with alginate can be adjusted to have different concentrations of the material, leading to altered flexibility. This capability has been used to print both stiff and soft models. In particular, the stiff hydrogel is used to print the root wall of the valve and soft hydrogel is used to print the leaflets [75]. Results showed that bioprinting can create mechanically heterogeneous anatomical heart valve conduits. Major cell types that make up the human heart valve are SMCs, valvular endothelial cells (VECs), and VICS [65, 69]. Most studies use one of these cell types to determine the viability of the cells within the construct post-printing over desired periods of time. For example, porcine aortic VICs were cultured for up to 21 days in a construct made of a PEG-DA hydrogel with nearly 100% viability [75]. More recent findings show that anatomically complex, heterogeneously seeded constructs can be created using 3D bioprinting. This was done, for instance, using an alginate/gelatin bioink, printed with directly encapsulated human SMCs in the root wall of the construct and porcine aortic VICs in the leaflets of the valve [74].

5.6 3D Bioprinted Stents

3D bioprinting has been used to fabricate stents for endovascular or coronary implantation to maintain vessel patency in partially or fully occluded vessels [76]. Coronary artery disease (CAD) is attributed to plaque buildup (atherosclerosis) within the arteries affecting blood supply to cardiac muscle by narrowing the vessel lumen. Current treatments for CAD include coronary artery bypass grafting (CABG) and angioplasty [77]. CABG is usually performed as a treatment option when there is complete blockage of the vessel lumen, where an artery is grafted around the blocked areas of the artery [78]. Angioplasty or balloon angioplasty is a less invasive option for partial blockages [79]. Biomaterials for 3D printed stents often include polymers such as polylactic acid (PLA) and polycaprolactone (PCL) while stents containing metal alloys such as stainless steel and cobalt chromium are still widely used [80, 81]. For instance, fused deposition modeling (FDM) has been used to efficiently print composite polymer stents [82]. FDM is a rapid prototyping process where a thermoplastic polymer is heated to melting point and then extruded in a layer by layer fashion to create a 3D model [82]. Implantable, biodegradable, and polymer-based stents have been produced for cardiovascular applications, exhibiting minimal toxicity and suitable degradation rate for tissue remodeling [81].

5.7 Summary and Concluding Remarks

In summary, 3D bioprinting shows much promise for the future of cardiac medicine [60]. It allows complexity in engineered models, specifically allowing the cardiac construct to be personally created based on the individual patient needs. It also allows for creating heterogeneous structures by using multiple extruders/inks for the prints, which is critical for recapitulating cardiovascular tissues with varying biochemical and physiomechanical properties (e.g., cellar composition and stiffness).

Bioprinted cardiac patches may not be fully viable yet as a clinical therapy for human patients with acute MI, until a functional vascular network is incorporated within the tissues. Co-culture integration and 3D fiber arrangement have been used to achieve some degrees of patch vascularization [34]. Other significant considerations must also be made, such as post-implantation cardiac arrhythmia occurrences among tested subjects and the patch's contractile capabilities. Implanted cardiac cells in nonhuman primates have been successfully shown to be able to remuscularize infarct primate but have been complicated by occurrences of arrhythmia [22]. These potential arrhythmic complications concomitant to cardiac patch implantation will need to be eliminated in order for patient applicability. As more advancements are made, though, with increased vascularization and little arrhythmogenicity, the plausibility of having cardiac patches available for clinical use becomes more likely.

One of the biggest setbacks with 3D bioprinting heart valves is that little testing has been done on the printed models under dynamic, physiologic conditions. Such conditions are essential for the accurate development of the extracellular matrix and tissue biomechanical testing [14, 83]. In addition, further testing is necessary to determine whether the valves can withstand the high pressures created by the ventricles during contraction, without tearing or regurgitation of blood upstream [72]. In the future, more studies will be needed to see whether these valves would be functional in an in vivo setting and if they would be a viable replacement for mechanical and biological valves.

References

1. Nakada T, Akiba T, Inagaki T, Morikawa T (2014) Thoracoscopic anatomical subsegmentectomy of the right S2b + S3 using a 3D printing model with rapid prototyping. Interact Cardiovasc Thorac Surg 19(4):696–698
2. Murphy SV, Atala A (2014) 3D bioprinting of tissues and organs. Nat Biotechnol 32(8):773–785
3. Wang Z, Lee SJ, Cheng HJ, Yoo JJ, Atala A (2018) 3D bioprinted functional and contractile cardiac tissue constructs. Acta Biomater 70:48–56
4. Cheung DYC, Duan B, Butcher JT (2015) Chapter 21—Bioprinting of cardiac tissues. In: Atala A, Yoo JJ (eds) Essentials of 3D biofabrication and translation. Academic, Boston, pp 351–370
5. Jang J (2017) 3D Bioprinting and in vitro cardiovascular tissue modeling. Bioengineering (Basel) 4(3):E71
6. Duan B (2017) State-of-the-art review of 3D bioprinting for cardiovascular tissue engineering. Ann Biomed Eng 45(1):195–209
7. Elshazly MB, Hoosien M (2018) Chapter 13—The future of 3D printing in cardiovascular disease. In: Al'Aref SJ, Mosadegh B, Dunham S, Min JK (eds) 3D printing applications in cardiovascular medicine. Academic, Boston, pp 243–253
8. Dunham S, Mosadegh B, Romito EA, Zgaren M (2018) Chapter 4—Applications of 3D printing. In: Al'Aref SJ, Mosadegh B, Dunham S, Min JK (eds) 3D printing applications in cardiovascular medicine. Academic, Boston, pp 61–78

9. Giannopoulos AA, Mitsouras D, Yoo S-J, Liu PP, Chatzizisis YS, Rybicki FJ (2016) Applications of 3D printing in cardiovascular diseases. Nat Rev Cardiol 13:701
10. Ozbolat IT, Peng W, Ozbolat V (2016) Application areas of 3D bioprinting. Drug Discov Today 21(8):1257–1271
11. Shafiee A, Atala A (2016) Printing technologies for medical applications. Trends Mol Med 22(3):254–265
12. Seol Y-J, Kang H-W, Lee SJ, Atala A, Yoo JJ (2014) Bioprinting technology and its applications. Eur J Cardiothorac Surg 46(3):342–348
13. Bakhtiar SM, Butt HA, Zeb S, Quddusi DM, Gul S, Dilshad E (2018) Chapter 10—3D printing technologies and their applications in biomedical science. In: Barh D, Azevedo V (eds) Omics technologies and bio-engineering. Academic, New York, pp 167–189
14. Lueders C, Jastram B, Hetzer R, Schwandt H (2014) Rapid manufacturing techniques for the tissue engineering of human heart valves. Eur J Cardiothorac Surg 46(4):593–601
15. Vunjak-Novakovic G, Tandon N, Godier A, Maidhof R, Marsano A, Martens TP, Radisic M (2009) Challenges in cardiac tissue engineering. Tissue Eng Part B Rev 16(2):169–187
16. Biglino G, Moharem-Elgamal S, Lee M, Tulloh R, Caputo M (2017) The perception of a three-dimensional-printed heart model from the perspective of different stakeholders: a complex case of truncus arteriosus. Front Pediatr 5:209
17. Zhang YS, Arneri A, Bersini S, Shin S-R, Zhu K, Goli-Malekabadi Z, Aleman J, Colosi C, Busignani F, Dell'Erba V, Bishop C, Shupe T, Demarchi D, Moretti M, Rasponi M, Dokmeci MR, Atala A, Khademhosseini A (2016) Bioprinting 3D microfibrous scaffolds for engineering endothelialized myocardium and heart-on-a-chip. Biomaterials 110:45–59
18. Holzl K, Lin S, Tytgat L, Van Vlierberghe S, Gu L, Ovsianikov A (2016) Bioink properties before, during and after 3D bioprinting. Biofabrication 8(3):032002
19. Gopinathan J, Noh I (2018) Recent trends in bioinks for 3D printing. Biomater Res 22:11
20. Serpooshan V, Zokaei S, Bagheri R (2007) Effect of rubber particle cavitation on the mechanical properties and deformation behavior of high-impact polystyrene. J Appl Polym Sci 104(2):1110–1117
21. Parikh NI, Gona P, Larson MG, Fox CS, Benjamin EJ, Murabito JM, O'Donnell CJ, Vasan RS, Levy D (2009) Long-term trends in myocardial infarction incidence and case fatality in the National Heart, Lung, and Blood Institute's Framingham Heart Study. Circulation 119(9):1203–1210
22. Zhang JY, Zhu WQ, Radisic M, Vunjak-Novakovic G (2018) Can we engineer a human cardiac patch for therapy? Circ Res 123(2):244–265
23. Serpooshan V, Hu JB, Chirikian O, Hu DA, Mahmoudi M, Wu SM (2018) Chapter 8—4D printing of actuating cardiac tissue. In: Al'Aref SJ, Mosadegh B, Dunham S, Min JK (eds) 3D printing applications in cardiovascular medicine. Academic, Boston, pp 153–162
24. Serpooshan V, Mahmoudi M, Hu DA, Hu JB, Wu SM (2017) Bioengineering cardiac constructs using 3D printing. J 3D Print Med 1(2):123–139
25. Mahmoudi M, Yu M, Serpooshan V, Wu JC, Langer R, Lee RT, Karp JM, Farokhzad OC (2017) Multiscale technologies for treatment of ischemic cardiomyopathy. Nat Nanotechnol 12(9):845–855
26. Zhu Y, Serpooshan V, Wu S, Demirci U, Chen P, Guven S (2017) TISSUE engineering of 3D organotypic microtissues by acoustic assembly. Methods Mol Biol https://www.ncbi.nlm.nih.gov/pubmed/28921421
27. Martins AM, Vunjak-Novakovic G, Reis RL (2014) The current status of IPS cells in cardiac research and their potential for tissue engineering and regenerative medicine. Stem Cell Rev Rep 10(2):177–190
28. Lee S, Serpooshan V, Tong X, Venkatraman S, Lee M, Lee J, Chirikian O, Wu JC, Wu SM, Yang F (2017) Contractile force generation by 3D hiPSC-derived cardiac tissues is enhanced by rapid establishment of cellular interconnection in matrix with muscle-mimicking stiffness. Biomaterials 131:111–120
29. Serpooshan V, Chen P, Wu H, Lee S, Sharma A, Hu DA, Venkatraman S, Ganesan AV, Usta OB, Yarmush M, Yang F, Wu JC, Demirci U, Wu SM (2017) Bioacoustic-enabled patterning of human iPSC-derived cardiomyocytes into 3D cardiac tissue. Biomaterials 131:47–57

30. Serpooshan V, Mahmoudi M (2015) Micropatterned nanostructures: a bioengineered approach to mass-produce functional myocardial grafts. Nanotechnology 26(6):060501
31. Wang QL, Wang HJ, Li ZH, Wang YL, Wu XP, Tan YZ (2017) Mesenchymal stem cell-loaded cardiac patch promotes epicardial activation and repair of the infarcted myocardium. J Cell Mol Med 21(9):1751–1766
32. Le Bras A (2018) Exosome-based therapy to repair the injured heart. Nat Rev Cardiol 15(7):382
33. Wang B, Borazjani A, Tahai M, Curry ALD, Simionescu DT, Guan JJ, To F, Elder SH, Liao J (2010) Fabrication of cardiac patch with decellularized porcine myocardial scaffold and bone marrow mononuclear cells. J Biomed Mater Res A 94a(4):1100–1110
34. Zhang J (2015) Engineered tissue patch for cardiac cell therapy. Curr Treat Options Cardiovasc Med 17(8):399
35. Serpooshan V, Wu SM (2014) Patching up broken hearts: cardiac cell therapy gets a bioengineered boost. Cell Stem Cell 15(6):671–673
36. Xin M, Olson EN, Bassel-Duby R (2013) Mending broken hearts: cardiac development as a basis for adult heart regeneration and repair. Nat Rev Mol Cell Biol 14(8):529–541
37. Ong CS, Fukunishi T, Zhang HT, Huang CY, Nashed A, Blazeski A, DiSilvestre D, Vricella L, Conte J, Tung L, Tomaselli GF, Hibino N (2017) Biomaterial-free three-dimensional bioprinting of cardiac tissue using human induced pluripotent stem cell derived cardiomyocytes. Sci Rep 7:4566
38. Wang Q, Yang H, Bai A, Jiang W, Li X, Wang X, Mao Y, Lu C, Qian R, Guo F, Ding T, Chen H, Chen S, Zhang J, Liu C, Sun N (2016) Functional engineered human cardiac patches prepared from nature's platform improve heart function after acute myocardial infarction. Biomaterials 105:52–65
39. Maiullari F, Costantini M, Milan M, Pace V, Chirivì M, Maiullari S, Rainer A, Baci D, Marei HE-S, Seliktar D, Gargioli C, Bearzi C, Rizzi R (2018) A multi-cellular 3D bioprinting approach for vascularized heart tissue engineering based on HUVECs and iPSC-derived cardiomyocytes. Sci Rep 8(1):13532
40. Hu JB, Hu DA, Buikema JW, Chirikian O, Venkatraman S, Serpooshan V, Wu SM (2017) Bioengineering of vascular myocardial tissue; a 3D bioprinting approach. Tissue Eng Part A 23:S158–S159
41. Hu JB, Tomov ML, Buikema JW, Chen C, Mahmoudi M, Wu SM, Serpooshan V (2018) Cardiovascular tissue bioprinting: physical and chemical processes. Appl Phys Rev 5(4):041106
42. Ong CS, Nam L, Ong K, Krishnan A, Huang CY, Fukunishi T, Hibino N (2018) 3D and 4D bioprinting of the myocardium: current approaches, challenges, and future prospects. Biomed Res Int 2018:11
43. Lux M, Andrée B, Horvath T, Nosko A, Manikowski D, Hilfiker-Kleiner D, Haverich A, Hilfiker A (2016) In vitro maturation of large-scale cardiac patches based on a perfusable starter matrix by cyclic mechanical stimulation. Acta Biomater 30:177–187
44. Gao L, Gregorich ZR, Zhu W, Mattapally S, Oduk Y, Lou X, Kannappan R, Borovjagin AV, Walcott GP, Pollard AE, Fast VG, Hu X, Lloyd SG, Ge Y, Zhang J (2018) Large cardiac muscle patches engineered from human induced-pluripotent stem cell-derived cardiac cells improve recovery from myocardial infarction in swine. Circulation 137(16):1712–1730
45. Randles A, Frakes DH, Leopold JA (2017) Computational fluid dynamics and additive manufacturing to diagnose and treat cardiovascular disease. Trends Biotechnol 35(11):1049–1061
46. Morris PD, Narracott A, von Tengg-Kobligk H, Silva Soto DA, Hsiao S, Lungu A, Evans P, Bressloff NW, Lawford PV, Hose DR, Gunn JP (2016) Computational fluid dynamics modelling in cardiovascular medicine. Heart 102(1):18–28
47. Sun Q, Groth A, Aach T (2012) Comprehensive validation of computational fluid dynamics simulations of in-vivo blood flow in patient-specific cerebral aneurysms. Med Phys 39(2):742–754
48. Ionita CN, Mokin M, Varble N, Bednarek DR, Xiang J, Snyder KV, Siddiqui AH, Levy EI, Meng H, Rudin S (2014) Challenges and limitations of patient-specific vascular phantom fabrication using 3D Polyjet printing. Proc SPIE Int Soc Opt Eng 9038:90380m

49. Maragiannis D, Jackson MS, Igo SR, Schutt RC, Connell P, Grande-Allen J, Barker CM, Chang SM, Reardon MJ, Zoghbi WA, Little SH (2015) Replicating patient-specific severe aortic valve stenosis with functional 3D modeling. Circ Cardiovasc Imaging 8(10):e003626
50. Gounley J, Chaudhury R, Vardhan M, Driscoll M, Pathangey G, Winarta K, Ryan J, Frakes D, Randles A (2016) Does the degree of coarctation of the aorta influence wall shear stress focal heterogeneity? In: 2016 38th annual international conference of the IEEE Engineering in Medicine and Biology Society (EMBC), pp. 3429–3432
51. Rutkowski DR, Reeder SB, Fernandez LA, Roldán-Alzate A (2018) Surgical planning for living donor liver transplant using 4D flow MRI, computational fluid dynamics and in vitro experiments. Comput Methods Biomech Biomed Eng Imaging Vis 6(5):545–555
52. Bulusu KV, Plesniak MW (2013) Secondary flow morphologies due to model stent-induced perturbations in a 180° curved tube during systolic deceleration. Exp Fluids 54(3):1493
53. Nair P, Chong BW, Indahlastari A, Ryan J, Workman C, Haithem Babiker M, Yadollahi Farsani H, Baccin CE, Frakes D (2016) Hemodynamic characterization of geometric cerebral aneurysm templates treated with embolic coils. J Biomech Eng 138(2):021011-1–021011-8
54. de Zélicourt D, Pekkan K, Kitajima H, Frakes D, Yoganathan AP (2005) Single-step stereolithography of complex anatomical models for optical flow measurements. J Biomech Eng 127(1):204–207
55. Saugel B, Holzapfel K, Stollfuss J, Schuster T, Phillip V, Schultheiss C, Schmid RM, Huber W (2011) Computed tomography to estimate cardiac preload and extravascular lung water. A retrospective analysis in critically ill patients. Scand J Traúma Resusc Emerg Med 19:31
56. Bane O, Shah SJ, Cuttica MJ, Collins JD, Selvaraj S, Chatterjee NR, Guetter C, Carr JC, Carroll TJ (2015) A non-invasive assessment of cardiopulmonary hemodynamics with MRI in pulmonary hypertension. Magn Reson Imaging 33(10):1224–1235
57. Rajiah P, Mak C, Dubinksy TJ, Dighe M (2011) Ultrasound of fetal cardiac anomalies. AJR Am J Roentgenol 197(4):W747–W760
58. Gindes L, Hegesh J, Weisz B, Gilboa Y, Achiron R (2009) Three and four dimensional ultrasound: a novel method for evaluating fetal cardiac anomalies. Prenat Diagn 29(7):645–653
59. Huang C, Zhou Y, Mao X, Tong J, Zhang L, Chen F, Hao Y (2017) Fusion of optical coherence tomography and angiography for numerical simulation of hemodynamics in bioresorbable stented coronary artery based on patient-specific model. Comput Assist Surg (Abingdon) 22(Suppl 1):127–134
60. Vukicevic M, Mosadegh B, Min JK, Little SH (2017) Cardiac 3D printing and its future directions. JACC Cardiovasc Imaging 10(2):171–184
61. Ford MD, Nikolov HN, Milner JS, Lownie SP, DeMont EM, Kalata W, Loth F, Holdsworth DW, Steinman DA (2008) PIV-measured versus CFD-predicted flow dynamics in anatomically realistic cerebral aneurysm models. J Biomech Eng 130(2):021015-1–021015-9
62. Molony D, Park J, Zhou L, Fleischer C, Sun HY, Hu X, Oshinski J, Samady H, Giddens DP, Rezvan A (2018) Bulk flow and near wall hemodynamics of the rabbit aortic arch: a 4D PC-MRI derived CFD study. J Biomech Eng 141(1):011003-011014
63. Ruedinger KL, Zhou H, Trampe B, Heiser T, Srinivasan S, Iruretagoyena JI, Roldán-Alzate A (2018) Modeling fetal cardiac anomalies from prenatal echocardiography with 3-dimensional printing and 4-dimensional flow magnetic resonance imaging. Circ Cardiovasc Imaging 11(9):e007705
64. Serpooshan V, Quinn TM, Muja N, Nazhat SN (2011) Characterization and modelling of a dense lamella formed during self-compression of fibrillar collagen gels: implications for biomimetic scaffolds. Soft Matter 7(6):2918–2926
65. Cui H, Miao S, Esworthy T, Zhou X, Lee S-j, Liu C, Yu Z-x, Fisher JP, Mohiuddin M, Zhang LG (2018) 3D bioprinting for cardiovascular regeneration and pharmacology. Adv Drug Deliv Rev 132:252–269
66. Cao Y, Gu C, Sun G, Yu S, Wang H, Yi D (2012) Quadruple valve replacement with mechanical valves: an 11-year follow-up study. Heart Surg Forum 15(3):E145–E149

67. Antoniou A, Harky A, Yap J, Lall K, Bashir M (2018) Bioprosthetic aortic valve replacement: a telltale from the young. Ann Transl Med 6(10):185
68. Bowdish ME, Kumar SR, Starnes VA (2016) The Ross procedure: an excellent option in the right hands. Ann Transl Med 4(23):471
69. Cheung DY, Duan B, Butcher JT (2015) Current progress in tissue engineering of heart valves: multiscale problems, multiscale solutions. Expert Opin Biol Ther 15(8):1155–1172
70. van der Valk DC, van der Ven CFT, Blaser MC, Grolman JM, Wu PJ, Fenton OS, Lee LH, Tibbitt MW, Andresen JL, Wen JR, Ha AH, Buffolo F, van Mil A, Bouten CVC, Body SC, Mooney DJ, Sluijter JPG, Aikawa M, Hjortnaes J, Langer R, Aikawa E (2018) Engineering a 3D-bioprinted model of human heart valve disease using nanoindentation-based biomechanics. Nanomaterials (Basel) 8(5):E296
71. Jana S, Lerman A (2015) Bioprinting a cardiac valve. Biotechnol Adv 33(8):1503–1521
72. Mosadegh B, Xiong G, Dunham S, Min JK (2015) Current progress in 3D printing for cardiovascular tissue engineering. Biomed Mater 10(3):034002
73. Scanlan AB, Nguyen AV, Ilina A, Lasso A, Cripe L, Jegatheeswaran A, Silvestro E, McGowan FX, Mascio CE, Fuller S, Spray TL, Cohen MS, Fichtinger G, Jolley MA (2018) Comparison of 3D echocardiogram-derived 3D printed valve models to molded models for simulated repair of pediatric atrioventricular valves. Pediatr Cardiol 39(3):538–547
74. Duan B, Hockaday LA, Kang KH, Butcher JT (2013) 3D bioprinting of heterogeneous aortic valve conduits with alginate/gelatin hydrogels. J Biomed Mater Res A 101(5):1255–1264
75. Hockaday LA, Kang KH, Colangelo NW, Cheung PY, Duan B, Malone E, Wu J, Girardi LN, Bonassar LJ, Lipson H, Chu CC, Butcher JT (2012) Rapid 3D printing of anatomically accurate and mechanically heterogeneous aortic valve hydrogel scaffolds. Biofabrication 4(3):035005
76. van Lith R, Baker E, Ware H, Yang J, Farsheed AC, Sun C, Ameer G (2016) 3D-printing strong high-resolution antioxidant bioresorbable vascular stents. Adv Mater Technol 1(9):1600138
77. Kandaswamy E, Zuo L (2018) Recent advances in treatment of coronary artery disease: role of science and technology. Int J Mol Sci 19(2):E424
78. Benjamin EJ, Blaha MJ, Chiuve SE, Cushman M, Das SR, Deo R, de Ferranti SD, Floyd J, Fornage M, Gillespie C, Isasi CR, Jimenez MC, Jordan LC, Judd SE, Lackland D, Lichtman JH, Lisabeth L, Liu S, Longenecker CT, Mackey RH, Matsushita K, Mozaffarian D, Mussolino ME, Nasir K, Neumar RW, Palaniappan L, Pandey DK, Thiagarajan RR, Reeves MJ, Ritchey M, Rodriguez CJ, Roth GA, Rosamond WD, Sasson C, Towfighi A, Tsao CW, Turner MB, Virani SS, Voeks JH, Willey JZ, Wilkins JT, Wu JH, Alger HM, Wong SS, Muntner P (2017) Heart disease and stroke statistics-2017 update: a report from the American Heart Association. Circulation 135(10):e146–e603
79. Dundar Y, Hill RA, Bakhai A, Dickson R, Walley T (2004) Angioplasty and stents in coronary artery disease: a systematic review and meta-analysis. Scand Cardiovasc J 38(4):200–210
80. Tappa K, Jammalamadaka U (2018) Novel biomaterials used in medical 3D printing techniques. J Funct Biomater 9(1):17
81. Guerra AJ, Cano P, Rabionet M, Puig T, Ciurana J (2018) 3D-printed PCL/PLA composite stents: towards a new solution to cardiovascular problems. Materials 11(9):E1679
82. Zein I, Hutmacher DW, Tan KC, Teoh SH (2002) Fused deposition modeling of novel scaffold architectures for tissue engineering applications. Biomaterials 23(4):1169–1185
83. Ziegelmueller JA, Zaenkert EK, Schams R, Lackermair S, Schmitz C, Reichart B, Sodian R (2010) Optical monitoring during bioreactor conditioning of tissue-engineered heart valves. ASAIO J 56(3):228–231

Chapter 6
Understanding Cancer Cell Behavior Through 3D Printed Bone Microenvironments

Yangyang Luo, Anusha Elumalai, Ahmed Humayun, and David K. Mills

Abstract Cancer is a significant health problem worldwide and forms through orchestrated and highly complex biological processes. This process is mediated through biophysical and biochemical signals that develop from within the tumor microenvironment. Although two-dimensional culture systems of established breast cancer cell lines are the most widely used model for cancer biology and preclinical drug assessments, it poorly mimics the behavior of cancer cells in vivo and fails to reproduce the in vivo tumor microenvironment, and accordingly, the data it produces is not always predictive. Effective therapeutic strategies require a cost-efficient in vitro model that can more accurately resemble the in vivo tumor microenvironment, thus permitting a variety of in vitro studies. Rapid prototyping (RP) is one of the most promising techniques for designing and producing three-dimensional (3D) systems (hydrogels, scaffolds) for drug efficacy analysis, developing drug delivery systems, and tissue engineering applications. The application of 3D bioprinting in engineering a cancer cell microenvironment will be the focus of this chapter. We will describe previous model systems used to understand cancer cell behavior. In particular, bioprinting methods and strategies that emphasize recreation of a cancer microenvironment that promotes cultured cancer cells to express a more relevant phenotype will be examined. Our focus will be on the 3D bioprinted models that serve as a predictive model for understanding mechanisms leading to cancer cell metastasis,

Y. Luo · A. Humayun
Molecular Science and Nanotechnology Program, Louisiana Tech University, Ruston, LA, USA
e-mail: ylu003@latech.edu; ahu005@latech.edu

A. Elumalai
Molecular Science and Nanotechnology Program, Louisiana Tech University, Ruston, LA, USA

School of Biological Sciences, Louisiana Tech University, Ruston, LA, USA

D. K. Mills (✉)
School of Biological Sciences, Louisiana Tech University, Ruston, LA, USA

Center for Biomedical Engineering and Rehabilitation Science, Louisiana Tech University, Ruston, LA, USA
e-mail: dkmills@latech.edu

M. Guvendiren (ed.), *3D Bioprinting in Medicine*,
https://doi.org/10.1007/978-3-030-23906-0_6

permit real-time study of cell–cell interactions, enable the analysis of growth factors and cytokine expression that supports tumor cell growth or survival, and the molecular cross-talk between tumor and stromal cells. The chapter will conclude with an assessment of the current state-of-the-art and future prospects.

Keywords Cancer · Culture models · Drug delivery · 3D bioprinting · Hydrogels · Metastasis · Microenvironment · Rapid prototyping · Nanoparticles

6.1 Introduction

Cancer can be defined as an uncontrolled cellular growth and spread of abnormal cells. It remains, worldwide, the second most common cause of death and it is predicted that by 2020 approximately 18 million new cases of cancer are expected [1, 2]. In the USA, there were 1,688,780 new cancer cases in 2017 and 600,920 cancer deaths are projected to occur [1]. The overall number of cancer deaths worldwide in 2008 was 7,564,802 [2, 3]. Cancer remains a significant cause of mortality for patients in the clinical setting. Patient death is strongly influenced by the cancer type, and 13% of the US population will die from this disease [4]. Among cancer types, the best survival rate (mortality/incidence × 100) is seen in patients with thyroid and testis cancer (16% and 18%, respectively), lung cancer (18.2%) while liver and pancreatic cancer have the highest mortality rates (93% and 96%, respectively). For all cancer types, the incidence of cancer is higher among men (20%) than in women, and the death rate among men is 40% higher [1]. The greatest fear among patients with cancer is pain, growing weakness, reduced quality of life, and medical costs [3, 4].

6.2 An Overview of Current Cancer Treatments

Cancer treatment is mainly based on surgery, radio- and chemotherapy but also other therapies are available: hyperthermia, targeted therapy, immuno- or phototherapy, the use of nanoparticles, stem cell transplant, or many lesser used therapies such as tumor ablation (Fig. 6.1) [5–8]. There also exists a critical need for drug delivery systems that are selective, targetable, deliver the proper dosage, and offer combinatorial delivery (in association with other agents or drugs).

Many cancer drugs cannot be delivered through commonly used administration routes because they are susceptible to degradation. Drug delivery systems have been under intensive investigation as a means to overcome this obstacle and achieve the desired therapeutic effect in humans or animals [9, 10]. Numerous drug delivery methods have been used over time, with the most common routes of administration being injectable, peroral, topical, or inhalation. The desired features of pharmaceutical carriers for parenteral administration include small size, biodegradability, high drug load, and prolonged circulation time. The most popular drug delivery systems

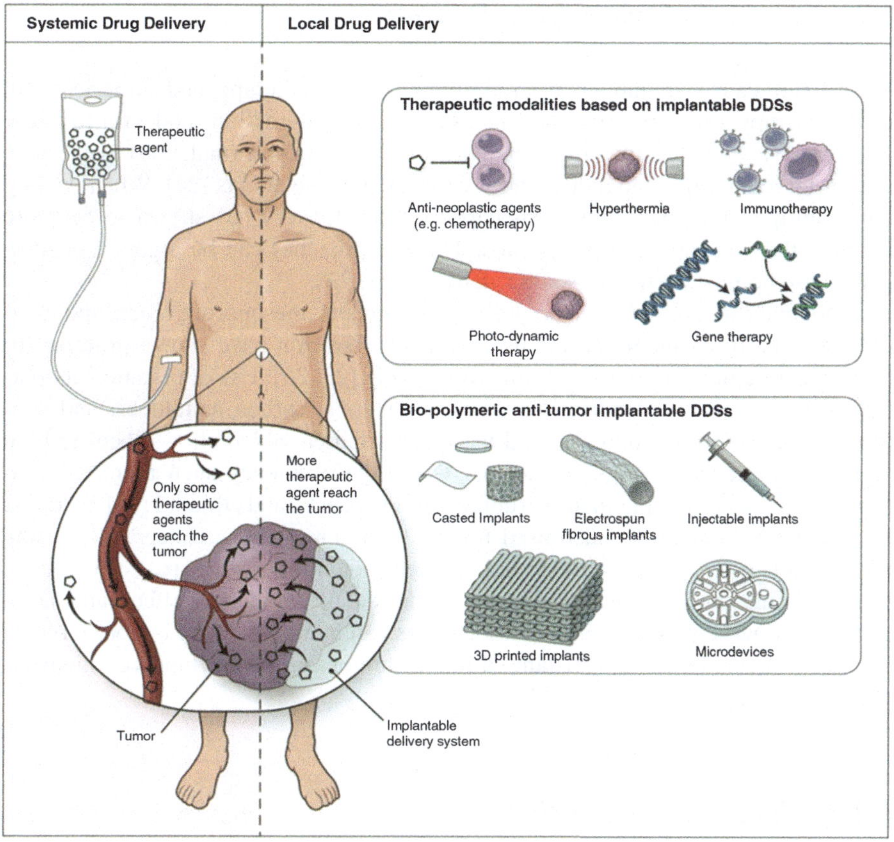

Fig. 6.1 Graphic illustrating modes of systemic drug delivery versus local drug delivery for cancer therapy. Note the various types of biopolymeric implantable systems used for local drug delivery. Many different anticancer therapeutic modalities can be employed using these implantable systems [20.]

are based on polymers, ceramics, and their composites [9]. By far, polymers have been used the most in developing drug delivery systems, with the most used being PEG, PEO, PCL, chitosan, alginate, polyvinyl alcohol, PMMA, etc. [11–14] Proteins, such as cellulose and collagen, have also been used as a matrix in some drug delivery systems [14]. Many different types and shapes have been prepared from albumin, soy and whey protein, collagen, and gelatin and used to deliver drug molecules using protein-based systems, including various protein cages, microspheres, nanoparticles, hydrogels, films, mini rods, and micropellets [15, 16]. Compared to synthetic polymers, their biocompatibility, biodegradability, low toxicity, cost-effectiveness, and availability make them well suited for drug delivery [17]. Various proteins-based systems have been developed including the ferritin/apoferritin protein cage, plant-derived viral capsids, and the small Heat shock protein (sHsp) cage [18, 19].

6.2.1 Systemic Cancer Drug Delivery

Mainstream cancer treatments use radiotherapy, chemotherapy, and surgeries [20]. Chemotherapy treats by reducing metastasis, cell progression, and proliferation. The drawback of this therapy is it unavoidably attacks normal body cells also. Tumor-targeted drugs can reduce the severity of adverse effects [21]. When dealing with multiple malignant sites, nanocarriers drug delivery has shown to be more efficient than chemotherapeutic drugs. These approaches can be organic, or inorganic based on the particles used for delivery [21, 22].

Inorganic carriers carbon nanotubes, quantum dots, mesoporous silica, and magnetic nanoparticles including the use of metals like iron have shown promise for traceable drug delivery with real-time monitoring [22, 23]. Use of organic nanoparticles like liposomes, dendrimers, lipids, emulsions, carbon nanotubes, and synthetic polymers have small size and large surface area which allows them to bind and carry anticancer drugs and agents with high efficiency such as many antitumoral drugs, e.g., cisplatin, doxorubicin, mitomycin C, and ethiodol [24]. Ceramic drug delivery systems are also used for the treatment of bone cancer [20]. Using nanoparticles for drug delivery has gained potential in the past few years. Encapsulating cancer drugs using nanoparticles decreases the side effects and based on the nanoparticle platform type, these drugs can be modified for a more targeted approach directed at drug-resistant strains of tumor cells with improved bioavailability and specificity [21, 22, 25].

6.2.2 Targeted Drug Delivery

Compared to the conventional drug therapies, targeted drug deliveries have a lot of advantages such as entering the tissue at a molecular level with increased localization and cellular uptake [26]. These drugs can reach out a larger surface via direct and selective targeting cancerous cells and have modifiable electronic, biologic, magnetic, and optical properties [27]. These drugs can identify molecular changes and pass through biological barriers while mediating molecular interactions. The greatest achievement these drugs have over conventional methods is the feasibility to recognize the cancerous cells by adding organic molecules, antibodies, peptides, and nucleic acids to the nanoparticles used in the drugs [27, 28]. Nanoparticles with PEG and modified PEG have a prolonged circulation time in the bloodstreams. Drugs, like albumin conjugated paclitaxel and liposomal doxorubicin are being clinically used and have exhibited significant efficacy. Natural polymers like dextran, chitosan, gelatin, hyaluronic acid, and collagen polypeptides are now being used as compoents of drug delivery systems for the treatment of cancerous tissues [28, 29].

PEG, PLGA, PVA, chitosan, PLA, PCL, and polyethylene oxide are biodegradable polymers used to prepare nanofibers for drug delivery. Hydrogels usually in the form of microspheres or nanoparticles allow for a more controlled drug release [30].

Current targeted drug therapies are focusing on using the addition of folate and transferrin receptors (Fig. 6.2) [31, 32]. Receptor-based targeted drugs like etoposide, gemcitabine, and vinorelbine have been developed and tested. The integration of nanoparticles along with therapeutic agents has created a new trend for targeted and precise drug release within cancerous sites by altering the conventional kinetics and distribution of cancer treatment.

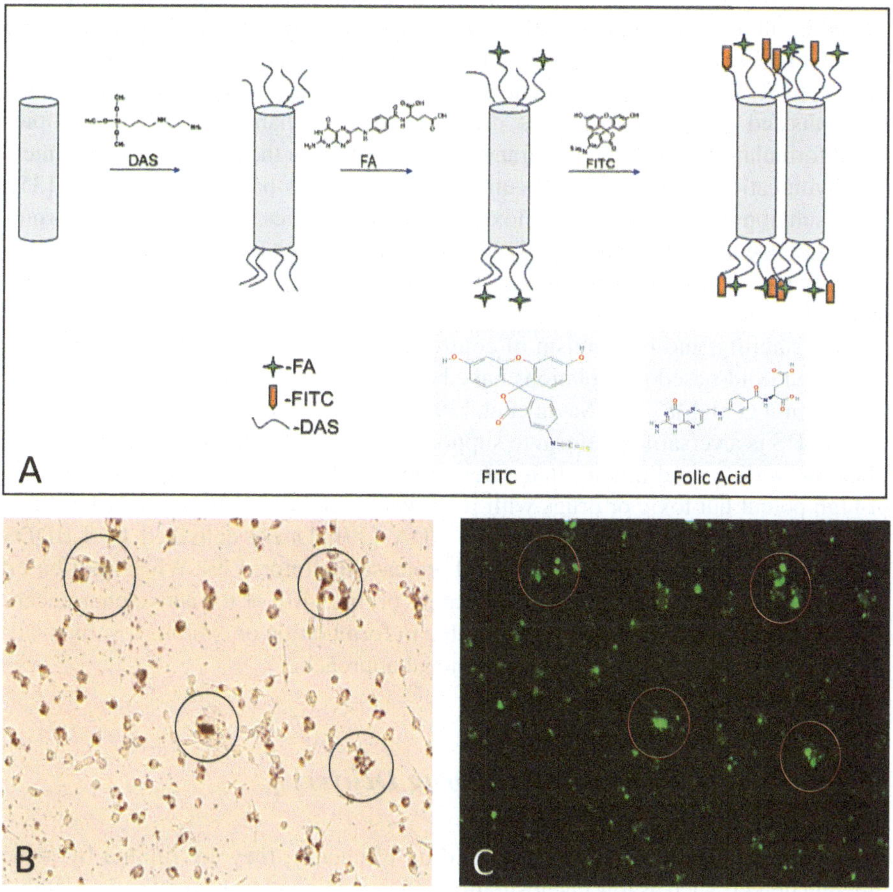

Fig. 6.2 (**a**) Schematic representation of the conjugation of both FA and FITC to DAS which is attached to the surface of the halloysite nanotubes (HNTs). (**b**) Phase contrast of bifunctionalized HNTS. (**c**) Same visual filed but viewed under fluorescent light. **Key:** DAS = *N*-[3-(trimethoxysilyl) propyl] ethylenediamine, FA = folic to the surface of the HNTs. DAS = *N*-[3-(trimethoxysilyl) propyl] ethylenediamine, FA = folic acid, FITC = fluorescein isothiocyanate. Micrographs courtesy of the author

6.2.3 Lipid-Based Drug Delivery Systems

Drugs that are applied through oral administration usually encounter problems of low permeability, rapid metabolism, and nonspecific targeting [33]. Furthermore, most of the newly developed anticancer drugs are highly hydrophobic (low solubility), toxic and exhibit short half-lives. Those drawbacks limit their use in treatment [34]. Lipid-based drug delivery systems (LBDDS) are one of the emerging technologies to address this problem. The lipid-based formulations convert drugs to be solubilized formulation. Simultaneously, the exogenous components of lipid-based drug formulation (surfactants, co-solvents, and complexation agent) induce changes of GI environment that increased secretion of biliary-derived solubilizing components (bile salts and phospholipids). Thus, drug solubility is enhanced [35]. Besides, lipid-based formulation alternates drug transport pathway from the portal vein to the intestinal lymphatic system, which circumvents hepatic first passing metabolism [35]. Encapsulation or solubilization of toxic chemotherapeutics with lipid-based formulation provide a safe and prolonged mean to deliver a drug to the intended site [35].

Lipid-based formulations can also be tailored to meet a wide range of requirements dictated by disease indication, selective absorption, root of administration, product stability, and mechanism of controlled release [34, 36]. Detailed guidance for design lipid-based formulations have been extensively reviewed by Porter et al. [37], Jannin et al. [38], and Savla et al. [39].

LBDDS is a versatile platform to support a diverse set of highly lipophilic drugs. They are widely used due to their safety, bioavailability, and efficacy for delivery of high potent but toxic or drugs with poor water-solubility. There are 36 lipid formulations that have been approved by FDA [39]. Drugs delivered by LBDDS include retinoids, vitamin D analogs, and protease inhibitors [39]. When lipid-based formulations cannot provide a suitable safety profile [40], or the active pharmaceutical ingredient (API) included, is not stable in formulation or exhibits low bioavailability [41], the LBDDS may not be the ideal approach.

6.2.4 Nanotechnology-Based Drug Delivery

Even though liposomes act to improve drug efficiency, they are limited by some inherent defects such as low encapsulation efficiency, rapid leakage of water-soluble drugs, and poor stability [42]. Nanoparticles (NPs) are another group of drug carriers that are used to improve the therapeutic index of drugs by improving drug solubility, reducing unwanted side effects, and delivering drugs to the target site (Fig. 6.3). Compared to liposomes, NPs have better stability [43] and offer controlled release [42].

Drug delivery nanoparticles are fabricated in the size range from 10 to 1000 nm. Usually, they are made of natural or artificial polymers such as polystyrene (PS), poly(methyl methacrylate) (PMMA), poly(lactic acid) PLA, and so on [43]. After

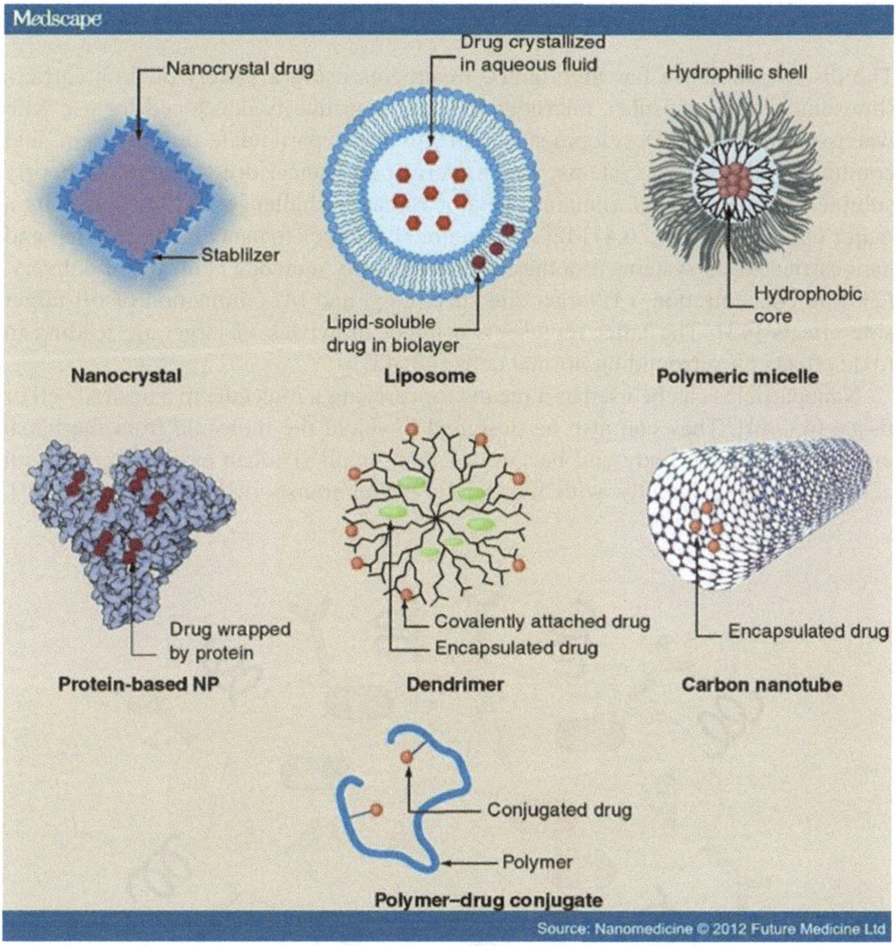

Fig. 6.3 Examples of nanotechnology-based drug delivery systems. Source: Nanomedicine © 2012 Future Medicine Ltd

decades of development, the materials used to make nanoparticle have increased. In a recent review, the types of nanoparticles used in drug delivery include not only polymers but also organometallic particles, viruses, and even liposomes [44].

Despite the significant advantages of NPs, several obstacles need to overcome to promote its clinical advancement. It has been found NPs would distribute and accumulate in the lung, liver, spleen, and kidney [45]. Also, biological barriers inhibit the accumulation of therapeutics tumor sites, which results in low tumor therapeutic efficiency [46]. To address those obstacles, smart extended release NPs (SER NPs) are developed. SER NPs are modified to release drugs with specific stimulation, such as heat, ultrasound, electric/magnetic fields, pH changes, and light [47]. The current advances of SER NPs are comprehensively reviewed by Kalaydina et al. [47].

6.2.4.1 Nanocontainers and Nanocarriers

The discussion above has highlighted treatments using a variety of drug carriers (liposomes, microcapsules, microbeads/spheres) primarily developed for use with water-soluble drugs. Development of effective nanoparticulate drug carriers, and combinatorial delivery systems, able to deliver anticancer drugs, including poorly soluble pharmaceuticals, remains a major research challenge and also presents a major opportunity (Fig. 6.4) [42]. The major challenges to using nanocontainer and nanocarrier-based systems as a therapeutic modality include (1) means of delivery, (2) drug concentration, (3) target site specificity, and (4) elimination of off-target side effects [43]. The latter represents a major drawback of targeting, leading to toxic effects on surrounding normal cells [42, 43].

Nanoparticles can be used as a means for carrying a molecule to a desired cell or tissue [43, 44]. They can also be designed to shield the molecule from the harsh environment of the body and because of their small size can easily interact with cells by binding directly with the cell or being transported into the cell [45].

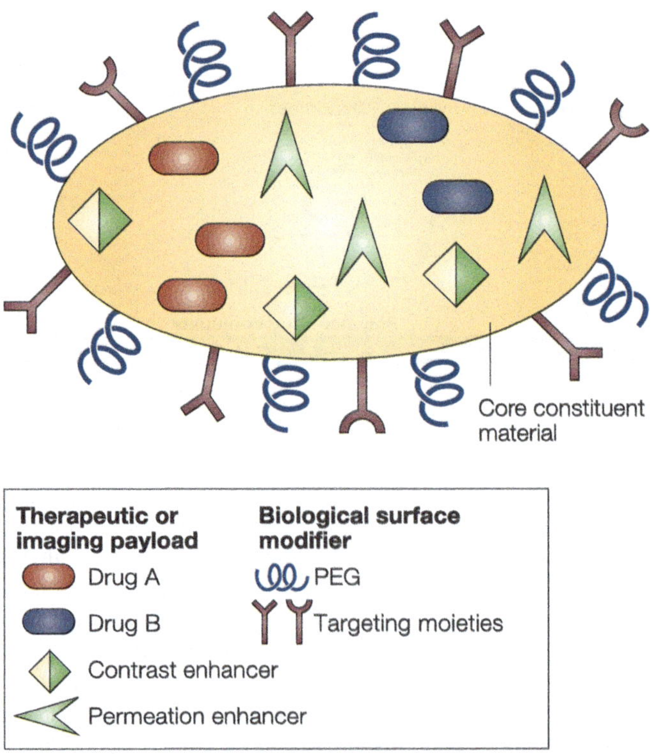

Fig. 6.4 Multifunctional nanoparticle example. In this system, one or more therapeutic agents can be delivered through specific targeting ligands. The ability to visualize the nanoparticle and avoid detection as well as open up biobarriers for delivery is highly desired [43]

Lipinski's rules severely limit the types of molecules that can currently be used to treat various diseases. By encapsulating molecules that would otherwise be destroyed in the body, inside of a nanocontainer, various anticancer agents can be used to the fight the disease and other disease types as well [43, 45].

6.2.4.2 Metal Nanoparticles

Cancer research has progressed by leaps and bounds; however, targeted cancer therapy that limits damage to non-cancer tissue remains the holy grail of cancer research. Anticancer medications paired with metal nanoparticles are a widely researched area in preclinical research. Metal nanoparticles (MNPs) having tunable physical properties with enhanced physical and chemical properties are a promising candidate for targeted specific cancer therapies (Fig. 6.5) [48]. Gold and silver nanoparticles are widely studied as potential therapeutic agents chiefly due to their strong surface plasmon resonance holding a possible key to the future of biosensors [49].

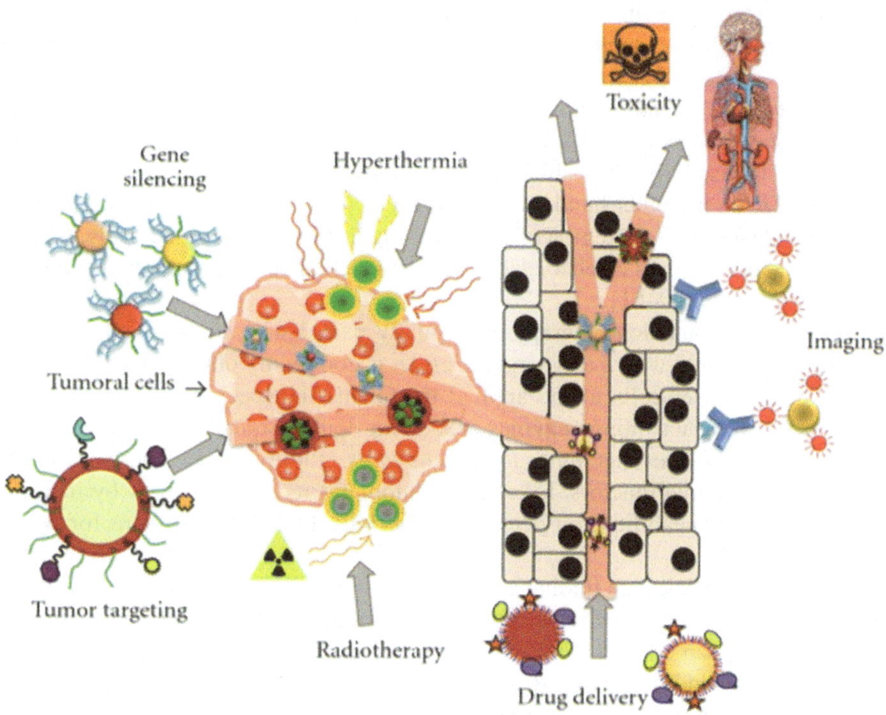

Fig. 6.5 MNPs can exploit characteristics of the newly formed vasculature supplying target tumors. MNPs can act simultaneously as therapeutic agents, inducing hyperthermia, enhancing radiotherapy, silencing genes, and/or delivering drugs to induce tumor cell death, and as imaging enhancers or contrast agents

MNPs can be used to target tumor tissue, passively or actively, and in some cases the buildup of MNPs due to inadequate waste removal in older tumor cells results in their bioaccumulation leading to cellular degradation. MNPs can be conjugated with antibodies, amino acids, peptides, and nucleotides to accomplish a more specific action and with conjugation of fluorescent tracker moieties make them an attractive anticancer therapy candidate [49].

6.2.4.2.1 Cancer Hyperthermal Therapy

Since the early nineteenth century, hyperthermal treatments (HT) involving localized heating of tumorous tissue have been used for therapy. HT is based on the principle that all living cells heated in the range of 41–48 °C exhibit apoptosis. Accordingly, rapidly dividing cancer cells are more susceptible to HT as they have poorly developed vasculature, hence they are more suceptible to heat as compared to healthy tissue. Furthermore, localized increases in temperature leads to reduced oxygen transport reducing oxygen levels for cancer cells while non-cancer cells remain relatively unaffected [49]. Higher temperatures can also induce necrosis due to denaturing of proteins leading to loss of enzyme function in mitochondrial-associated release of cytochrome C. HT further triggers heat shock proteins that can activate T cells and generate an immune response. HT can be local, regional, or whole body wherein only a subset of tumorous tissue, an entire organ or a body cavity and the whole body is subjected to HT, respectively. Localized HT delivers extremely focused thermal ablation using radio, ultrasound, and microwaves, regional HT, on the other hand, is based on selective isolation and heating of systemic blood supply to the region of interest using heating devices, and lastly, whole-body hyperthermia uses electric blankets, thermal chambers, or water baths to raise body temperature [49–51].

HT is not a standalone therapy and usually paired with chemo- and radiation therapy respectively where a synergistic effect is commonly observed due to sensitizing of the cancer cells to the applied treatment. HT specifically complements radiotherapy by inactivating enzymes needed for double-strand DNA repair, affects S-phase cells, and prevents tumor angiogenesis by inhibiting tumor-derived vascular endothelial growth factors. Likewise, it supplements chemotherapy by increasing cell wall permeability and decreasing the interstitial fluid pressure for better drug uptake [52].

Metal nanoparticles are finding extensive use in localized and targeted heating of tissues using gold nanoparticles, iron oxide nanoparticles, and carbon nanotubes conjugates that can convert electromagnetic energy into thermal energy, thus proving a particular and controlled therapy. Gold silica nanoparticle conjugate having plasmon absorption in the near infrared spectrum has been used as photothermal agents to convert electromagnetic radiation to thermal energy. However, they are limited to only tumors located near the body surface, whereas tumors situated far from the body surface pose a challenge due to dissipation and absorption of the near infrared radiation by the healthy body tissue [52].

6.2.4.2.2 Cancer Magnetic Therapy

While HT is a promising method for conjoint anticancer treatment, it suffers from several limitations including nonspecific heating and damage to non-tumor tissue sites; in this regard, cancer magnetic therapy where in magnetic energy is converted into thermal energy may be a feasible alternative. Cancer magnetic therapy uses magnetic nanoparticles (MNPs) such as iron oxide core in the form of magnetite [Fe_3O_4] or maghemite [γFe_2O_3] that can be encapsulated within a polymeric hydrophilic shell for increased stability and biocompatibility; they can also be conjugated with other metal nanoparticles such as gold to facilitate the attachment of functional moieties such as antibodies and antigens. MNPs can be guided under magnetic fields to the area of interest where on the application of an oscillating magnetic field they can generate heat through friction and oscillation. Depending on the particle size and intensity of magnetic field, they have a distinct advantage over HT using a magnetic field they can be guided to a region which could easily be a deep-seated tumor inaccessible to infrared radiation or localized heating [53]. Similarly, drugs can be adsorbed to MNPs and guided to therapy site using magnetic fields, thus reducing/regulating the systemic circulation of the drug. They can also be conducted using implants, e.g., intrathecal implants in the spinal cord for treating central nervous system tumors; a preclinical study reported that using MNPs the chemotherapeutic drug could be localized in the tumor and using only 5–10% dose a 57.2% remission could be achieved [53, 54]. Further MNPs conjugated with antibodies, adjuvants, or antigens to elicit an immune response towards tumorous tissue in cancer immunotherapy is a promising field of research [54, 55].

6.3 Cancer Metastasis

Cancer cells can metastasize and migrate to other organs of the body. Breast cancer, in particular, is often prone to metastasis to other organs, resulting in a highly problematic condition that is often associated with high morbidity [56, 57]. Prostate, lung, melanoma, and skin cancers also preferentially home to the bone tissues and the frequency of bone metastasis is as high as ~70% in these cancer types [58]. Advanced bone metastasis accounts for the high morbidity in breast cancer patients. For example, the 5-year survival rate of breast cancer patients after bone metastasis is only 26% [57] and breast cancer is the second leading cause of cancer deaths in women wherein approximately 85% of individuals will eventually develop bone metastases [58, 59].

6.3.1 Bone as a Microenvironment for Cancer Metastasis

As a cancerous tumor grows, cells break away and are carried to other parts of the body by the blood or lymphatic system [60]. One of the most common places for tumor cells to migrate is to the bone stroma, especially with breast, prostate, kidney, thyroid, and lung cancer [56, 59, 60]. The most common bones invaded by cancer cells are the femur, humerus, spine, ribs, pelvis, and the skull. After tumor cells have occupied bone and proliferated, they develop into tumor masses; they are frequently associated with severe symptoms, including bone pain, leucoerythroblastic anemia, bone deformity, nerve-compression syndromes, hypercalcemia, and high levels of calcium within blood, that often causes nausea, fatigue, thirst, and frequent urination [59, 60]. Bone weakening due to tumor growth often leads to pathological fractures [61, 62].

The interaction between cancer cells and the host microenvironment is considered to be responsible for the establishment of metastatic lesions [62]. Cancer cells homing to the bone might not immediately form metastases but remain in bone marrow as solitary dormant cells to evade apoptosis induced by factors in a foreign microenvironment [61, 62]. Colonies of cancer cells can reside in bone tissue as quiescent cells for years before they become aggressive and grow into macrometastases. These dormant cancer cells do not divide and thus have been observed clinically to be resistant to chemotherapy. Some quiescent cancer cells at some point can switch to a proliferative phenotype that is more aggressive in nature [61]. However, there is relatively little known about the underlying mechanisms controlling cell cycle regulation and dormancy of solitary metastatic cells. The interaction between cancer cells and bone tissue during this long time period is currently not well understood. The prognosis for bone metastasis remains abysmal with few treatment options available. Successful treatment depends on detection of metastases and bone invasion at its earliest stages and understanding the metastatic cascade at the cellular level as a prelude to developing new drugs and therapeutics [61, 63].

Despite the advances in the field of cancer treatment, increased knowledge of the etiology and disease progression of cancer, the introduction of new cancer drugs, and advances in prevention, and early detection, cancer continues to be a significant health concern [56, 58, 59]. Breast cancer treatment could be significantly improved if three key problems were considered. Cancer treatments should be personalized to each patient, not a "one size fits all" approach. A more detailed understanding of the mechanisms of drug resistance and relapse must be understood [58]. Newer drugs and novel treatment modalities (co-administration, synergetic drugs, and drug delivery system) should be identified, developed, and tested [58, 59]. Fundamental research into the underlying mechanisms of cancer causation is also critical in addressing these issues. Experimentation in the laboratory at the cellular level provides key information; however, much depends on the design, development, and use of appropriate in vitro models and robust assays to deliver accurate and representative cellular and molecular data [59, 60].

6.4 Current Treatments for Bone Cancer Metastasis

Treatments for bone metastases are designed to stop or slow tumor cell growth, preventing further bone damage [61]. Bone-modifying drugs, which are medications that slow bone thinning, reduce pain, and decrease hypercalcemia, may be given [62]. A bone metastasis that is located only in one area is commonly treated with radiation therapy to relieve pain and strengthen the bone. Surgery may be used to remove a tumor or prevent or treat a bone fracture, and a special cement can be injected into a bone to stabilize it [64]. Chemotherapy, hormone therapy, and the use of radioactive drugs are treatment options if bone metastases are found in more than a single area [65, 66].

6.5 In Vitro and In Vivo Models of the Cancer Microenvironment

6.5.1 Two-Dimensional (2D) Monolayer vs. Three-Dimensional (3D) Culture Systems

Much of cancer research relies on animal models; but these have associated feasibility and ethical concerns [67, 68]. These issues may be overcome using in vitro models that allow for a more controlled approach to investigating cancer cell biology, compound screening, drug efficacy studies, and disease modelling. The vast majority of such models involve culturing cells on conventional two-dimensional (2D) cell culture dishes in which cells adapt to the flat substrate, flatten, grow as monolayers and may not fully express their 3D in vivo phenotype. Unfortunately, this approach is a poor substitute for animal models and does not mimic the environment that cells experience in vivo [69, 70]. Although 2D cell culture affords simplicity and low cost, it is generally acknowledged that it fails to adequately reproduce the tumor environment. The morphology of cultured cells in monolayers cultures is changes, cell-to-cell contact is limited, and the microenvironment generated by the cells in monolayer offers biological, chemical, and mechanical cues that are reduced or altered [71, 72].

To study cancer metastasis, three-dimensional (3D) in vitro models are becoming a popular alternative to 2D culture assays and animal models. 3D systems represent a more realistic approach, enabling cells to retain their native 3D morphology, form representative interactions with adjacent cells, and create more complex ECM structures [70–72]. Many scaffolds have been fabricated over the past 20 years using various compositions and geometry to study tumor cell growth [73, 74]. These scaffolds serve as mimicked microenvironments to model and define mechanisms associated with cancer cell behavior [75]. These 3D matrices are traditionally composed of natural extracellular matrix (ECM) gels and synthetic polymers.

A major obstacle to understanding the cancer microenvironment in vitro is the lack of realistic bone tissue model suitable for long-term study of cell-to-cell interactions. An in vitro bone metastasis model is ideal for various mechanistic studies; however, the existing in vitro models are unable to fully replicate the cancer microenvironment [67, 68]. Current culture models failed to mimic the bone tissue, as it is a complex tissue where living cells are embedded in a mineralized organic matrix composed mostly of hydroxyapatite, Type I collagen, and bone-specific proteins. Recently, functional 3D bone tissues have been developed by combining osteogenic cells with artificial scaffolds such as synthetic polymers, titanium, and ceramics [76]. Compared to 2D models, these 3D models make significant improvements in cell differentiation, mineral deposition, tissue organization, and physiological response to different stimuli. Nevertheless, a natural bone microenvironment composed of cell-synthesized, mineralized ECM cannot be achieved by simply mixing cells with synthetic scaffolds [77].

6.5.2 Cancer Spheroids

Many biofabrication techniques have been developed for building tissues organs as well as smaller organoids or spheroids [78, 79]. In contrast, models using cellular spheroids offer a natural 3D microenvironment, where human cancer cells or small tissue fragments can be transplanted into an animal to facilitate cellular growth and metastasis. These approaches have several distinct advantages. It can facilitate rapid tissue formation and accelerate tissue maturation [80, 81]. 3D in vitro models have become a popular alternative that overcomes the limitations inherent in 2D culture assays and animal models.

As a cancer model, in vitro 3D models allow a controlled approach to studying cancer cell biology. These systems allow cells to grow and function and produce a more representative tissue model. They further enable the creation of a morphologically and physiologically realistic microenvironment allowing a study of cellular behavior, drug delivery, and drug discovery [82]. 3D cancer models permit cultured cancer cells to form interactions with adjacent cells, the creation of tissues with a high cell density, culture of multiple cell types, and production of more complex structures [81, 83]. The 3D microenvironment often consists of highly porous and inert scaffolds with multiple voids and interconnections wherein cells can freely grow and occupy, creating cell appropriate ECMs in vitro [84].

Microfluidic 3D tissue models recreate complex in vivo microenvironment including scale, morphology, hemodynamics, and cellular interactions. 3D systems represent a more realistic approach that can be coupled with real-time visualization, study of cell-to-cell communication, and analysis of secreted cellular products [85]. The microfluidic technology can provide micro-scale complex structures and well-controlled parameters to mimic the in vivo environment of cells. The combination of microfluidic technology and 3D cell culture has seen the development of many in vivo-like tissue-based applications, specific cancer models, drug cancer efficacy and screening, and organ-on-a-chip systems [86, 87]. For an excellent review, please see Duinen et al. [88].

6.6 3D Bioprinting

The application of 3D printing in medicine can provide many benefits, including: the customization and personalization of medical products, drugs, and equipment; cost-effectiveness; increased productivity; the democratization of design and manufacturing; and enhanced collaboration [89, 90]. Medical uses for 3D printing, both actual and potential, can be organized into several broad categories, including: tissue and organ fabrication; creation of customized prosthetics, implants, and anatomical models; and pharmaceutical research regarding drug dosage forms, delivery, and discovery.

3D cell bioprinting is a relatively new bioengineering tool being used to create 3D cell constructs for regenerative medicine and transplantation therapies [91]. A broad range of printing technologies have been developed and are being used to deliver cells and biomaterials for disease remediation, tissue repair and regeneration, and organ transplantation [91–93]. The building block of this technology is the "bioink," typically a hydrogel-based composite of cells, instructive agents, and polymers used in the printing process to produce three-dimensional cell structures into designed architectures to generate small tissues or organs [94, 95].

Organ transplantation is the established treatment modality for patients whose vital organs are failing (or failed) such as the kidney, pancreas, liver, heart, or lung [91, 96–98]. Restoration of organ function is vital to ensure the affected patient does not die from a fatal disease or vital organ failure [94]. The need for organs replacement is a great challenge in clinical medicine. In particular, the demand for 3D printed organs and tissues is high owing to a shortage of donor organs, rising rates of lifestyle diseases (i.e., type II diabetes), trauma (gunshots, disasters, armed conflicts), and technological innovations such as autonomously driven vehicles (reduction in donated organs) [95, 99, 100]. Accordingly, there is an intense research effort focused on the application of 3D printing for the production of biomedical materials and devices for dental and orthopedic applications. Medical applications for 3D printing are expanding rapidly and are expected to revolutionize health care [94–98, 100].

6.6.1 3D Bioprinting Devices

3D bioprinting offers additional important advantages beyond the traditional regenerative method, which essentially provides seeded cells in a scaffold for support [91, 97]. The technical advantages of 3D bioprinting include highly precise cell placement and high digital speed control, resolution, cell concentration, drop volume, diameter of printed cells and inclusion of instructive bioactive agents, i.e., growth factors [92, 100, 101].

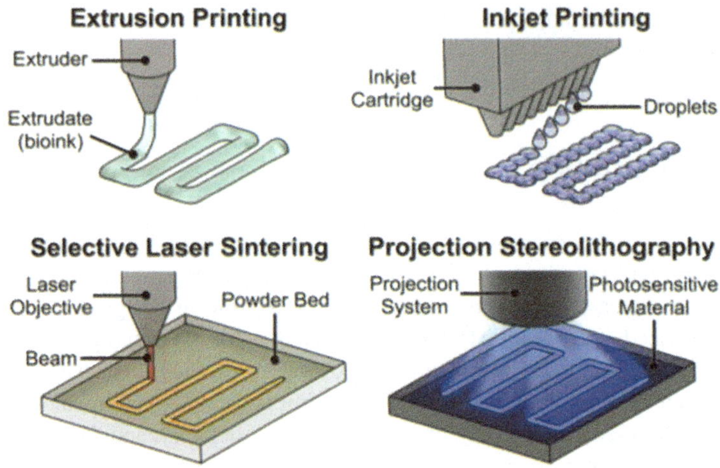

Fig. 6.6 3D printing methods commonly adapted for biomaterials fabrication. Extrusion printing and inkjet printing rely on liquid intermediates or precursors which can solidify quickly after ejection. Selective laser sintering provides localized heating to melt or fuse powder granules. Stereolithography relies on photoinduced polymerization of a liquid resin in the specific regions exposed to light. (Artwork by Jacob Albritton and Jordan Miller)

Bioprinting of biological relevant materials has significant advantages over the previous approaches as it allows for the precise deposition of cells, biomaterials, and bioactive cues in three-dimensional space that is a better mimic for the native tissue (Fig. 6.6) [92, 93]. Many different bioprinting systems exist; they can be laser-based, inkjet-based, or extrusion-based, where inkjet-based bioprinting is most common [94, 97]. This method deposits a "bioink," droplets of living cells or biomaterials, onto a substrate according to a computer-designed program to reproduce human tissues or organs [95]. Bioinks have specific requirements (and constraints) such as its viscosity as this depends on the printing process.

6.6.1.1 Inkjet-Based Bioprinting

Inkjet printing is one of the oldest printing methods and it still holds promise for 3D printing of biological and biomedical applications [95]. This method is also known as drop-on-demand printing, drop-by-drop or drop-on-demand bioprinting [99, 100]. Inkjet printing is a noncontact reprographic strategy and it is based on the deposition of bioink droplets that are natural or man-made materials [101]. The objective is to mimic an extracellular matrix environment and support the cell adhesion, proliferation, and differentiation of mammalian cells [102].

The fabrication strategy for producing cell-seeded droplets can be printed and assembled into a construct in a layer-by-layer fashion [103]. Inkjet printing relies on three different mechanisms including piezoelectric inkjet (acoustic) [103, 104], thermal inkjet [99, 103], and electrostatic bioprinting [105]. Piezoelectric bioprinters generate acoustic waves within the bioink chamber by using a piezoelectric actuator that ejects droplets through the printer nozzle [104]. The thermal method contains a fluid chamber and single, double, or multiple nozzle heads. Within the bioink chamber, heat is generated and induces pulses of pressure, and the buildup pressure results in the ejection of picoliter volume of droplets at the nozzle orifice [8, 99, 103]. In electrostatic bioprinters, droplets are generated by voltage pulses which are applied between a pressure plate and an electrode [15, 106].

These strategies also have some disadvantages for use in tissue biofabrication especially the use of high viscose materials, high cell density loading can cause clogging of the nozzle, droplet bioprinters use low viscosity bioinks, and the mechanical strength of printed structures is generally inferior when compared to the target tissue [107–109].

6.6.1.2 Laser-Based Bioprinting

Laser-based bioprinting or biological laser printing is a method for that patterns arrays of living cells for tissue biofabrication. This strategy of printing is based on a long wavelength laser or high energy light source [110, 111]. Cells are printed by laser beam pulsating at controlled rates [112] onto a receiving substrate. A typical laser bioprinting device set-up consists of a (1) pulsed laser beam, (2) a focusing system, (3) a laser absorbing ribbon, (4) a receiving substrate, and (5) a cell-containing material [113, 114]. Several factors influence the print outcome including the resolution of laser bioprinting, surface tension, the air gap between the substrate and the ribbon, wettability of the substrate, thickness and viscosity of the organic layer, and laser type and its configuration [115]. Unlike the limitations with inkjet printing, laser-based bioprinting can print a range of material viscosities (1–300 mPa/s) [20]. The many advantages of laser bioprinting is that it is nozzle-free so factors such as high density and viscosity are not an issue. Laser bioprinting also has the ability to printing mammalian cells without significant adverse effects on cell viability and functionality. However, this approach is not a commonly used method but laser-based bioprinting is seeing increased application in tissue and organ-engineering applications [95].

6.6.1.3 Extrusion-Based Printers

Extrusion bioprinting, also known as direct writing, is one of the more commonly used 3D printing methods [116]. In extrusion bioprinting, the bioink is dispensed by mechanical force typically via a screw or piston or by pneumatic gas or pressurized air. The bioink is extruded continuously as a strand. Extrusion bioprinting generally consists of dispenser (single ejector or multiple ejector) systems that are placed on an automated robotic stage and controlled by a stage controller. The robotic scene has three axes (x–y–z) [117]. The bioink containing encapsulated cells is deposited on a building substrate placed directly below the dispenser [118]. The major advantages of this type of printing method are its compatibility for printing materials with an extensive range of viscosities, bioinks with high cell densities, specialty hydrogels, and biodegradable thermoplastics such as polycaprolactone [26, 119]. Additionally, it is a rapid method with an acceptable cell viability post-printing but also has a low fabrication resolution (~200 μm) [25, 118].

For an extrusion-based bioprinting approach, the bioink should show shear thinning behavior to facilitate extrusion through the printer needle and maintain cell viability [94, 95]. Further, the ink needs to display quick shear recovery behavior to guarantee immediate termination of flow upon deposition on a substrate in order to retain the desired shape. A material with these properties allows the fabrication of large structures with high resolution. Cross-linking or printing into a supporting matrix (the fresh system) also permits shape retention [120]. Apart from the strong material requirements during the printing process, the final mechanical properties of the printed construct should also be sufficient for long-term shape fidelity, easy handling, and post-printing processing [121]. Lastly, the bioink also needs to be viable, extrudable and provide a supportive microenvironment [120].

6.6.1.4 Bioplotting

Bioplotting is another 3D printing method that is used for in tissue engineering and biofabrication. Because of its versatility, it offers many interesting and new opportunities for biofunctional rapid prototyping [121]. Bioplotting uses a syringe to extrude either tubes or spheroids of materials. These kinds of printers usually have several syringes and can employ multiple cell types and fabrication conditions [119]. The result of using several syringes enables construction of multiple tissue types in the final construct, an ideal capability for producing bioengineered soft tissues [121]. In this system, layers are arranged on top of each other and cured through a chemical reaction or UV radiation [100].

Despite these advantages, the major challenge in bioplotting is choosing the materials for extrusion because the materials have to be viscous, cell supportive and provide a functional cellular microenvironment [122]. Biomaterials such as thermoset resins, polymer melts, pastes with high filler contents, cements, polymer solutions, and even macromolecules such as proteins can be used with this method [28, 123]. According to some research, bioplotting is the best methods for generating co-cultured scaffolds and tissues which don't require high resolution details [29].

6.6.1.5 Fused-Deposition Modeling (FDM)

FDM is the oldest 3D additive manufacturing technology used in rapid prototyping, modeling, and production applications [30]. This approach plays a significant role in improving dimensional precision, the quality of products, reduced production times, and cost of biofabrication [31]. This method of printing deposits a melted thermoplastic in thin layers that provides support for hydrogel-based bioinks. This process begins with a CAD model. The model is produced by melting materials into liquid state in a liquefier head to form layers of selectively deposited materials through a nozzle [30].

There are several parameters that can affect the resolution of FDM constructs like nozzle diameter and the type of polymers used. For having hard filament have to work into raw materials and there are a small minority of materials can employed for FDM technology [32]. Optimization of process parameters is one of the major important design tasks in FDM. Usually for obtaining high quality structure, the main research area is directed towards improving surface roughness, mechanical properties, material behavior, building time, and dimensional accuracy [30]. One of the major limitations of this method is lack of mechanical strength of molten thermoplastic because it cannot support itself during slow cooling and hardening [31, 32].

6.6.1.6 Stereolithography (SL)

Stereolithography was introduced in the late 1980s. It is a solid freeform technique which uses light to cross-link polymeric materials [33, 34]. Most stereolithography techniques use a laser (commonly UV light) and a directed mirror array to project a light beam onto the surface of a liquid photocurable resin. After the resin is cured, the fabrication platform moves in the z-direction, and then a fresh layer of resin is added, and the process repeated to build up a 3D construct. Stereolithography has become an accepted method because of its ability to prepare structures with high resolution and the ability to remove uncured resin from the final product. In spite of these advantages, this approach is slow and the resins that are used are non-biomimetic and biomaterials scientists are working to develop resins that are appropriate for use in tissue engineering applications [29]. Visible light-based stereolithography is a rapidly developing approach in bioprinting in which visible light crosslinkable bioinks have been introduced [35–37].

3D bioprinting also offers a robust tool to precisely arrange multiple cell types and biomaterials within one scaffold, leading to creating an organ containing multiple scaffolds and cell types [101]. 3D printing addresses many of the difficulties associated with the fabrication of artificial ECM scaffolds in its ability to control pore size, number, and distribution [92, 97, 101]. This technique has been used to print multiple tissues and organs in a variety of regenerative medicine applications [101, 102]. The precisely controlled geometry of the printed construct makes this technique preferable to conventional scaffold manufacturing strategies [102].

6.7 3D Printed Bone Cancer Microenvironment

Organ printing takes advantage of 3D printing technology to produce cells, biomaterials, and cell-laden biomaterials individually or in tandem, layer by layer, creating 3D tissue-like structures [92, 94, 101]. Various materials are available to build the scaffolds, and their selection is dependent on the desired strength, porosity, and tissue type. Hydrogels are generally considered to be most suitable for producing soft and hard tissues [100, 102]. A process for bioprinting organs has emerged. This includes the creation of an organ blueprint with its vascular architecture. This is followed by the bioprinting process design plan. Next, cell isolation and cellular cures are incorporated as a means to preserve cellular phenotype and with stem cells and to direct cellular differentiation [105, 106].

With the biocomponent completed, bioink reservoirs with organ-specific and blood cells are added in a supportive medium. These are then loaded into the printer and bioprinted. Maintenance, histogenesis, and organ formation will then occur within a bioreactor prior to transplantation [100–103].

A 3D printer with multiple print heads can be used to deposit different cell types (bone, cartilage muscle cells), a necessary feature for fabricating whole tissues and organs [105, 106]. Such knowledge enables the 3D printing cells and materials into tissues designed to provide a better understanding of how tumors are generated and how cancer cell behave [107]. Scaffold porosity, which is the percent of empty space created by pores in a structure, along with the size and shape of pores, can change the permeability of nutrients and media, impact cell attachment, and even facilitate cell migration. In an effort to simulate the cancer microenvironment, a number of 3D printed matrices have been reported [108, 109].

6.7.1 3D Printed Cancer Model Examples

6.7.1.1 Preserving Cancer Cell Behavior

Preservation of cancer cell behavior is crucial and identification of the optimal bioink (hydrogel) with cell supportive properties that promotes cancer cell behavior phenotype and behavior is a major research goal [124]. Zhao et al. developed a cervical tumor model by extruding HeLa cells and gelatin/alginate/fibrinogen hydrogels through a commercial syringe in a layer-by-layer fashion [125]. The authors found that increased cell proliferation, spheroid formation, higher protein expression and chemoresistance occurred in the 3D printed and patterned model when compared with 2D cell culture. In another study, ovarian cancer cells were co-cultured with fibroblasts and micropatterned simultaneously on Matrigel using an inkjet printer [126]. In the latter study, the presence of both cell types maintained viability and proliferation in the 3D matrix, suggesting the promise of this methodology. 3D printed models are increasing exploiting co-cultures and mixed cell cultures to understand cell-to-cell interactions [127–129].

Such models are a significant improvement over 2D model systems which provide not only a 3D microenvironment that supports cell growth, but also allow a means to study cell–cell interactions leading to cancer migration, progression, and metastasis.

6.7.1.2 Fabricating Bone Cancer Cell Models

Disease models for bone cancer, or metastatic cancers that spread to bone hold great promise in obtaining an understanding of how cancer and bone cells interact with each other. 3D printed cancer models of human tumors should mimic the structure, function, and drug response of the in vivo condition [130–132]. Bioprinting can facilitate this by fabricating 3D constructs that mimic tumor heterogeneity, vasculature, cell-to-cell interactions, and cell clusters and spheroids [133, 134]. The bioprinted models are enabling researchers in understanding tumor development, cellular cross-talk leading to cancer-induced bone disease (osteolysis osteoblastic disease), and the role of the vasculature and signaling agents [77, 124, 128, 129]. Zhou et al. (2016) used stereolithography to fabricate a 3D biomimetic bone matrix incorporating the integration of human fetal osteoblasts, human bone marrow MSCs, and BrCa cells to create a bone-like microenvironment for the study of cancer metastasis [128, 129]. For those readers seeking more information, please see the review by Bray et al. (2015) for a great review on this topic [130, 131].

Furthermore, bioprinting can be used to fabricate physiologically relevant microenvironments within a microfluidic device, thus creating "tumor-on-a-chip" mechanism for high-throughput testing in a biomimetic microenvironment [134, 135]. Applications of tumors-on-a-chip include facilitating basic research to better understand tumor development, structure, and function as well as drug screening to improve the efficiency of cancer drug discovery [135–137].

6.8 Current State-of-the-Art and Future Prospects

Over the last 10 years, cancer research has been an area marked by rapid progress and our understanding of cancer cell behavior has grown substantially. In particular, due to 3D culture models and now 3D bioprinted cancer microenvironments, we have reached a better understanding of cancer cell biology and the mechanisms of tumor formation and metastasis. 3D bioprinted cancer models hold the prospect of rapidly advancing our knowledge of cancer risk factors, cancer and host tissue cellular interactions, and factors leading to metastasis.

Understanding cancer biology and translating this knowledge to the clinic will improve the cancer therapy significantly. In particular, further understanding of the cancer stem cells in tumor biology, the cancer stem cell microenvironment, and the consequences of their inherent ability for self-renewal and the differentiation factors will be critical in developing new treatment modalities. Personalized treatment has the potential to increase treatment efficacy, and hence decrease mortality rates. Moreover, 3D printed microfluidic systems and lab-on-a-chip technology are promising vehicles for developing personalized oncology. With the ability to predict response or resistance to drug therapy, we can identify the right drug or drug combination that will be more effective, thus sparing patients the morbidity and impactful side effects.

Acknowledgements Support for this work was provided by the Center for Dental, Oral & Craniofacial Tissue & Organ Regeneration (C-DOCTOR) with the support of NIH NIDCR (U24DE026914).

References

1. Siegel RL, Miller KD, Jemal A (2017) Cancer statistics. CA Cancer J Clin 67:7–30
2. Jemal A, Bray F, Center MM, Ferlay J, Ward E, Forman D (2011) Global cancer statistics. CA Cancer J Clin 61:69–90. [PubMed: 21296855]
3. Luengo-Fernandez R, Leal J, Gray A et al (2013) Economic burden of cancer across the European Union: a population-based cost analysis. Lancet Oncol 14:1165–1174
4. Mariotto AB, Yabroff KR, Shao Y et al (2011) Projections of the cost of cancer care in the United States: 2010–2020. J Natl Cancer Inst 103:117–128
5. https://www.cancer.org/research/cancer-facts-statistics/all-cancer-facts-figures/cancer-facts-figures-2018.html. Last accessed 10 Aug 2018
6. DeSantis C, Siegel R, Bandi P, Jemal A (2011) Breast cancer statistics. CA Cancer J Clin 61:409–418. [PubMed: 21969133]
7. O'Shaughnessy J (2005) Extending survival with chemotherapy in metastatic breast cancer. Oncologist 10(Suppl 3):20–29. [PubMed: 16368868]
8. DePalma M, Hanahan D (2012) The biology of personalized cancer medicine: facing individual complexities underlying hallmark capabilities. Mol Oncol 6(2):111–127
9. Lawler M, Alsina D, Adams RA (2018) Critical research gaps and recommendations to inform research prioritization for more effective prevention and improved outcomes in colorectal cancer. Gut 67:179–193
10. Lord CJ, Ashworth A (2010) Biology-driven cancer drug development: back to the future. BMC Biol 8:38
11. Ghosh S, Ghosh S (2004) Recent research and development in synthetic polymer-based drug delivery systems. J Chem Res 4:241–246
12. Apicella A, Cappello B, Delnobile MA et al (1994) Poly(ethylene oxide)-based delivery systems. Polym Drugs Drug Adm 545:111–125
13. Liechty WB, Kryscio DR, Slaughter BV, Peppas NA (2010) Polymers for drug delivery systems. Annu Rev Chem Biomol Eng 1:149–173
14. Han HD, Song CK, Park YS et al (2008) A chitosan hydrogel-based cancer drug delivery system exhibits synergistic antitumor effects by combining with a vaccinia viral vaccine. Int J Pharm 350:27–3415
15. Verma D, Gulati N, Kaul S, Mukherjee S, Nagaich U (2018) Protein based nanostructures for drug delivery. J Pharm (Cairo) 2018:9285854. https://doi.org/10.1155/2018/9285854, 18 pages
16. Jao D, Xue Y, Medina J, Hu X (2017) Protein-based drug-delivery materials. Materials 10(5):517. https://doi.org/10.3390/ma10050517
17. Fuchs S, Coester C (2010) Protein-based nanoparticles as a drug delivery system: chances, risks, perspectives. J Drug Deliv Sci Technol 20(5):331–342
18. Yamazoe H (2018) Multifunctional protein microparticles for medical applications. Biomaterials 155(1):1–12
19. MaHam A, Tang Z, Wu Z, Wang J, Lin Y (2009) Protein-based nanomedicine platforms for drug delivery. Small 5(15):1706–1721
20. Talebian S, Foroughi J, Wade SJ, Vine KL, Dolatshahi-Pirouz A, Mehrali M, Conde J, Wallace G (2018) Biopolymers for antitumor implantable drug delivery systems: recent advances and future outlook. Adv Mater 30(31):e1706665. https://doi.org/10.1002/adma.201706665
21. Kawaski ES, Player A (2005) Nanotechnology, nanomedicine, and the development of new, effective therapies for cancer. Nanomedicine 1:101–109

22. Chakraborty C, Pal S, Doss GP, Wen ZH, Lin CS (2013) Nanoparticles as 'smart' pharmaceutical delivery. Front Biosci (Landmark Ed) 1(18):1030–1050
23. Fioramonti Calixto GM, Bernegossi J, Marise de Freitas L, Carla Raquel Fontana CR, Chorilli M (2016) Nanotechnology-based drug delivery systems for photodynamic therapy of cancer: a review. Molecules 21:342. https://doi.org/10.3390/molecules21030342
24. Wolinsky JB, Colson YL, Grinstaff MW (2012) Local drug delivery strategies for cancer treatment: gels, nanoparticles, polymeric films, rods, and wafers. J Control Release 159:14–26
25. Cafaggi S, Russo E, Stefani R, Leardi R, Caviglioli G, Parodi B et al (2007) Preparation and evaluation of nanoparticles made of chitosan or *N*-trimethyl chitosan and a cisplatin-alginate complex. J Control Release 121:110–123
26. Mishra N, Pant P, Porwal A, Jaiswal J, Samad M, Tiwari S (2016) Targeted drug delivery: a review. Am J PharmTech Res 6(1):1–24
27. Singh R, Lillard JW (2009) Nanoparticle-based targeted drug delivery. Exp Mol Pathol 6(3):215–223
28. Ruiz ME, Gantner ME, Talevi A (2014) Applications of nanosystems to anticancer drug therapy (Part II. Dendrimers, micelles, lipid-based nanosystems). Recent Pat Anticancer Drug Discov 9(1):99–128
29. Katsogiannou M, Peng L, Catapano CV, Rocchi P (2011) Active-targeted nanotherapy strategies for prostate cancer. Curr Cancer Drug Targets 11(8):954–965
30. Aguilar ZP (2013) Targeted drug delivery, Chap. 5. In: Nanomaterials for medical applications. Elsevier, New York, pp 181–234, ISBN 9780123850898
31. Tang MF, Lei L, Guo SR, Huang WL (2010) Recent progress in nanotechnology for cancer therapy. Chin J Cancer 29(9):775–780
32. Grimes R, Luo YY, Mills DK (2018) Bifunctionalized clay nanoparticles for cancer therapy. Appl Sci 8:281. https://doi.org/10.3390/app8020281
33. Kalepu S, Manthina M, Padavala V (2013) Oral lipid-based drug delivery systems—an overview. Acta Pharm Sin B 3(6):361–372
34. Yingchoncharoen P, Kalinowski DS, Richardson DR (2016) Lipid-based drug delivery systems in cancer therapy: what is available and what is yet to come. Pharmacol Rev 68(3):701–787
35. Porter CJ, Trevaskis NL, Charman WN (2007) Lipids and lipid-based formulations: optimizing the oral delivery of lipophilic drugs. Nat Rev Drug Discov 6(3):231
36. Shrestha H, Bala R, Arora S (2014) Lipid-based drug delivery systems. J Pharm (Cairo) 2014:801820
37. Porter CJ, Pouton CW, Cuine JF, Charman WN (2008) Enhancing intestinal drug solubilisation using lipid-based delivery systems. Adv Drug Deliv Rev 60(6):673–691
38. Jannin V, Musakhanian J, Marchaud D (2008) Approaches for the development of solid and semi-solid lipid-based formulations. Adv Drug Deliv Rev 60(6):734–746
39. Savla R, Browne J, Plassat V, Wasan KM, Wasan EK (2017) Review and analysis of FDA approved drugs using lipid-based formulations. Drug Dev Ind Pharm 43(11):1743–1758
40. Lu C, Perez-Soler R, Piperdi B, Walsh GL, Swisher SG, Smythe WR, Shin HJ, Ro JY, Feng L, Truong M et al (2005) Phase II study of a liposome-entrapped cis-platin analog (L-NDDP) administered intrapleurally and pathologic response rates in patients with malignant pleural mesothelioma. J Clin Oncol 23:3495–3501
41. Lemke A, Kiderlen AF, Kayser O (2005) Amphotericin B. Appl Microbiol Biotechnol 68:151–162
42. Soppimath KS, Aminabhavi TM, Kulkarni AR, Rudzinski WE (2001) Biodegradable polymeric nanoparticles as drug delivery devices. J Control Release 70(1–2):1–20
43. Mauro FM (2005) Cancer nanotechnology: opportunities and challenges. Nat Rev 5:151–171
44. Bamrungsap S, Zilong Z, Chen T, Wang L, Li C, Fu T, Tan W (2012) A focus on nanoparticles as a drug delivery system. Nanomedicine 7(8):1253–1271
45. Blanco E, Shen H, Ferrari M (2015) Principles of nanoparticle design for overcoming biological barriers to drug delivery. Nat Biotechnol 33(9):941

46. Pérez-Herrero E, Fernández-Medarde A (2015) Advanced targeted therapies in cancer: drug nanocarriers, the future of chemotherapy. Eur J Pharm Biopharm 93:52–79
47. Kalaydina RV, Bajwa K, Qorri B, Decarlo A, Szewczuk MR (2018) Recent advances in "smart" delivery systems for extended drug release in cancer therapy. Int J Nanomedicine 13:4727
48. Mody VV, Siwale R, Singh A, Mody HR (2010) Introduction to metallic nanoparticles. J Pharm Bioallied Sci 2(4):282–289
49. Conde J, Doria G, Baptista P (2012) Noble metal nanoparticles applications in cancer. J Drug Deliv 2012:751075. https://doi.org/10.1155/2012/751075, 12 pages
50. Gerweck LE (1985) Hyperthermia in cancer therapy: the biological basis and unresolved questions. Cancer Res 45(8):3408–3414
51. Wust P et al (2002) Hyperthermia in combined treatment of cancer. Lancet Oncol 3(8):487–497
52. Haranaka K, Sakurai A, Satomi N (1987) Antitumor activity of recombinant human tumor necrosis factor in combination with hyperthermia, chemotherapy, or immunotherapy. J Biol Response Mod 6(4):379–391
53. Chen J et al (2007) Immuno gold nanocages with tailored optical properties for targeted photothermal destruction of cancer cells. Nano Lett 7(5):1318–1322
54. Tietze R, Lyer S, Dürr S et al (2013) Efficient drug-delivery using magnetic nanoparticles—biodistribution and therapeutic effects in tumour bearing rabbits. Nanomedicine 9:961–971
55. Evans ER et al (2018) Metallic nanoparticles for cancer immunotherapy. Mater Today 21(6):673–685
56. Guan X (2015) Cancer metastases: challenges and opportunities. Acta Pharm Sin B 5(5):402–418
57. Jiang WG, Sanders AJ, Katoh M, Ungefroren U, Gieseler F, Prince M, Thompson SK, Zollo M, Spano D (2015) Tissue invasion and metastasis: molecular, biological and clinical perspectives. Semin Cancer Biol 35:S244–S275
58. Macedo F, Ladeira K, Pinho F, Saraiva N, Bonito N, Pinto L, Goncalves F (2017) Bone metastases: an overview. Oncol Rev 11(1):321. https://doi.org/10.4081/oncol.2017.321
59. Croucher PI, McDonald MM, Martin TJ (2016) Bone metastasis: the importance of the neighbourhood. Nat Rev Cancer 16:373–386
60. Gdowski AS, Ranjan A, Vishwanatha JK (2017) Current concepts in bone metastasis, contemporary therapeutic strategies and ongoing clinical trials. J Exp Clin Cancer Res 36:108. https://doi.org/10.1186/s13046-017-0578-1
61. Smith MA, Gurney JG et al (2017) Malignant bone tumors. Cancer incidence and survival among children and adolescents: United States SEER Program 1975-1995. National Cancer Institute, SEER Program, Bethesda.: NIH Pub No. 99-4649, pp 99–110
62. Steeg PS (2016) Targeting metastasis. Nat Rev Cancer 16:201–218
63. Redig AJ, McAllister SS (2013) Breast cancer as a systemic disease: a view of metastasis. J Intern Med 274(2):113–126. https://doi.org/10.1111/joim.12084
64. Bissell MJ, Radisky D (2001) Putting tumours in context. Nat Rev Cancer 1:46–54
65. Massagué J, Obenauf AC (2016) Metastatic colonization by circulating tumour cells. Nature 529:298–306
66. https://www.oncolink.org/cancers/bone/bone-metastases/bone-metastasis-treatment-with-medications. Last accessed 16 Jan 2019
67. Yamada KM, Cukierman E (2007) Modeling tissue morphogenesis and cancer in 3D. Cell 130(4):601–610
68. Kimlin LC, Casagrande G, Virador VM (2013) In vitro three-dimensional (3D) models in cancer research: an update. Mol Carcinog 52(3):167–182. https://doi.org/10.1002/mc.21844. Epub 2011 Dec 7
69. Alemany-Ribes M, Semino CE (2014) Bioengineering 3D environments for cancer models. Adv Drug Deliv Rev 79–80:40–49
70. Friedrich J, Ebner R, Kunz-Schughart L (2007) Experimental anti-tumor therapy in 3-D: spheroids--old hat or new challenge? Int J Radiat Biol 83:849–871

71. Fessart D, Begueret H et al (2013) Three-dimensional culture model to distinguish normal from malignant human bronchial epithelial cells. Eur Respir J 42(5):1345–1356. https://doi.org/10.1183/09031936.00118812. Epub 2013 Jan 24
72. Luca AC, Mersch S et al (2013) Impact of the 3D microenvironment on phenotype, gene expression, and EGFR inhibition of colorectal cancer cell lines. PLoS One 8(3):e59689
73. Cosson S, Otte EA, Hezaveh H, Cooper-White JJ (2015) Concise review: tailoring bioengineered scaffolds for stem cell applications in issue engineering and regenerative medicine. Stem Cells Transl Med 4:156–164
74. Knowlton S et al (2015) Bioprinting for cancer research. Trends Biotechnol 33:504–513
75. Baker BM, Chen CS (2012) Deconstructing the third dimension: how 3D culture microenvironments alter cellular cues. J Cell Sci 125(Pt 13):3015–3024
76. Zhou X, Zhu W, Nowicki M, Miao S, Cui H, Holmes B, Glazer RJ, Zhang LG (2016) 3D bioprinting a cell-laden bone matrix for breast cancer metastasis study. ACS Appl Mater Interfaces 844:30017–30026. https://doi.org/10.1021/acsami.6b10673
77. Sitarski AM, Fairfield H, Falank C, Reagan MR (2018) 3D tissue engineered in vitro models of cancer in bone. ACS Biomater Sci Eng 4(2):324–336. https://doi.org/10.1021/acsbiomaterials.7b00097
78. Mironov V, Boland T, Trusk T, Forgacs G, Markwald RR (2003) Organ printing: computer-aided jet-based 3D tissue engineering. Trends Biotechnol 21:157–161
79. Mironov V, Kasyanov V, Drake C, Markwald RR (2008) Organ printing: promises and challenges. Regen Med 3:93–103
80. Busse A, Letsch A et al (2013) Characterization of small spheres derived from various solid tumor cell lines: are they suitable targets for T cells? Clin Exp Metastasis 30(6):781–791. https://doi.org/10.1007/s10585-013-9578-5. Epub 2013 Mar 22
81. Perche F, Torchilin VP (2012) Cancer cell spheroids as a model to evaluate chemotherapy protocols. Cancer Bio Ther 13(12):1205–1213
82. Ingram M et al (2010) Tissue engineered tumor models. Biotech Histochem 85:213–229. https://doi.org/10.3109/10520295.2010.483655
83. Upreti M et al (2011) Tumorendothelial cell threedimensional spheroids: new aspects to enhance radiation and drug therapeutics. Transl Oncol 4:365–376
84. Chang R, Emami K, Hand W, Sun W (2010) Biofabrication of a three-dimensional liver micro-organ as an in vitro drug metabolism model. Biofabrication 2:045004
85. Mengus C, Manuele G, Muraro MG, Mele V, Amicarella F, Manfredonia C, Muenst S et al (2018) In vitro modeling of tumor–immune system interaction. ACS Biomater Sci Eng 4(2):314–323. https://doi.org/10.1021/acsbiomaterials.7b00077
86. Li XJ, Valadez AV, Zuo P, Nie Z (2102) Microfluidic 3D cell culture: potential application for tissue-based bioassays. Bioanalysis 4(12):1509–1525
87. Ghaemmaghami AM, Hancock MJ, Harrington H et al (2012) Biomimetic tissues on a chip for drug discovery. Drug Discov Today 17(3–4):173–181
88. Duinen V, Trietsch SJ, Joore J, Vulto P, Hankemeier T (2015) Microfluidic 3D cell culture: from tools to tissue models. Curr Opin Biotechnol 35:118–126
89. Mills DK (2015) Future medicine: the impact of 3D printing. J Nanomater Mol Nanotechnol 4(3):1–3. https://doi.org/10.4172/2324-8777.1000163
90. Ventola CL (2014) Medical applications for 3D printing: current and projected uses. Pharm Ther 39(10):704–711
91. Mironov V, Derby B (2006) Bioprinting: a beginning. Tissue Eng 12(4):631–639
92. Kang HW, Lee SJ, Ko IK, Kengla C, Yoo JJ, Atala AA (2016) 3D bioprinting system to produce human-scale tissue constructs with structural integrity. Nat Biotechnol 34:312
93. Hung BP, Naved BA, Nyberg EL, Dias M, Holmes CA, Elisseeff JJ et al (2016) Three-dimensional printing of bone extracellular matrix for craniofacial regeneration. ACS Biomater Sci Eng 2:1806–1804
94. Ji S, Guvendiren M (2017) Recent advances in bioink design for 3D bioprinting of tissues and organs. Front Bioeng Biotechnol 5:23–31

95. Murphy SV, Atala A (2014) 3D bioprinting of tissues and organs. Nat Biotechnol 32:773–785
96. Khatiwala C, Law R, Shepard B, Dorfman S, Ceste M (2012) 3D cell bioprinting for regenerative medicine research and therapies. Gene Ther Regul 7(1):1230004. (19 pages). https://doi.org/10.1142/S1568558611000301
97. Schubert C, van Langeveld MC, Donoso LA (2012) Innovations in 3D printing: a 3D overview from optics to organs. Br J Ophthalmol 98(2):159–161
98. Ozbolat IT, Yu Y (2013) Bioprinting toward organ fabrication: challenges and future trends. IEEE Trans Biomed Eng 60:691–699
99. Cu X, Boland T, D'Lima DD, Lotz MK (2012) Thermal inkjet printing in tissue engineering and regenerative medicine. Recent Pat Drug Deliv Formul 6(2):149–155
100. O'Brien CM, Holmes B, Faucett S, Zhang LG (2015) Three-dimensional printing of nanomaterial scaffolds for complex tissue regeneration. Tissue Eng Part B Rev 21:103
101. Gurkan UA et al (2014) Engineering anisotropic biomimetic fibrocartilage microenvironment by bioprinting mesenchymal stem cells in nanoliter gel droplets. Mol Pharm 11(7):2151–2159
102. Xu F, Sridharan B, Wang S, Gurkan UA, Syverud B, Demirci U (2011) Embryonic stem cell bioprinting for uniform and controlled size embryoid body formation. Biomicrofluidics 5(2):22207
103. Gudapati H, Dey M, Ozbolat I (2016) A comprehensive review on droplet-based bioprinting: past, present and future. Biomaterials 102:20–42
104. Wijshoff H (2010) The dynamics of the piezo inkjet printhead operation. Phys Rep 491(4):77–177
105. Tasoglu S, Demirci U (2013) Bioprinting for stem cell research. Trends Biotechnol 31(1):10–19
106. Kamisuki S, Hagata T, Tezuka C, Nose Y, Fujii M, Atobe M (1998) A low power, small, electrostatically-driven commercial inkjet head. In: Proceedings of the eleventh annual international workshop on Micro Electro Mechanical Systems, 1998. MEMS 98, pp 63–68
107. Guillotin B, Guillemot F (2011) Cell patterning technologies for organotypic tissue fabrication. Trends Biotechnol 29(4):183–190
108. Guillotin B et al (2010) Laser assisted bioprinting of engineered tissue with high cell density and microscale organization. Biomaterials 31:7250–7256
109. Sears NA, Seshadri DR, Dhavalikar PS, Cosgriff-Hernandez E (2016) A review of three-dimensional printing in tissue engineering. Tissue Eng Part B Rev 22(4):298–310
110. Sears NA, NA SDR, Dhavalikar PS, Cosgriff-Hernandez E (2016) A review of three-dimensional printing in tissue engineering. Tissue Eng Part B Rev 22(4):298–310
111. Nahmias Y, Odde DJ (2006) Micropatterning of living cells by laser-guided direct writing: application to fabrication of hepatic-endothelial sinusoid-like structures. Nat Protoc 1(5):2288
112. Koch L, Gruene M, Unger C, Chichkov B (2011) Laser assisted cell printing. Curr Pharm Biotechnol 14(1):91–97
113. Saunders R, Bosworth L, Gough J, Derby B, Reis N (2004) Selective cell delivery for 3D tissue culture and engineering. Eur Cell Mater 7(Suppl 1):84–85
114. Demirci U, Montesano G (2007) Single cell epitaxy by acoustic picolitre droplets. Lab Chip 7(9):1139–1145
115. Guillemot F, Souquet A, Catros S, Guillotin B (2010) Laser-assisted cell printing: principle, physical parameters versus cell fate and perspectives in tissue engineering. Nanomedicine 5(3):507–515
116. Mir TA, Nakamura M (2017) Three-dimensional bioprinting: toward the era of manufacturing human organs as spare parts for healthcare and medicine. Tissue Eng Part B Rev 23(3):245–256. https://doi.org/10.1089/ten.TEB.2016.0398. Epub 2017 Mar 21
117. Ferris CJ, Gilmore KG, Wallace GG (2013) Biofabrication: an overview of the approaches used for printing of living cells. Appl Microbiol Biotechnol 97(10):4243–4258
118. Arslan-Yildiz A, El Assal R, Chen P, Guven S, Inci F, Demirci U (2016) Towards artificial tissue models: past, present, and future of 3D bioprinting. Biofabrication 8(1):14103

119. Chartrain NA, Williams CB, Whittington AR (2016) Engineering tissues with bioprinting. BioProcess Int 14(10):2016
120. Hospodiuk M, Dey M, Sosnoski D, Ozbolat IT (2017) The bioink: a comprehensive review on bioprintable materials. Biotechnol Adv 35(2):217–239
121. Rutz L, Lewis PL, Shah RN (2017) Toward next-generation bioinks: tuning material properties pre- and post-printing to optimize cell viability. MRS Bull 42(8):563–570
122. Yang E, Miao S, Zhong J, Zhang Z, Mills DK, Grace Zhang LG (2018) Bio-based polymers for 3D printing of bioscaffolds. Poly Rev (Phila Pa) 58(4):668–687. https://doi.org/10.1080/15583724.2018.1484761
123. Abdelaal OAM, Darwish SMH (2013) Review of rapid prototyping techniques for tissue engineering scaffolds fabrication. In: Öchsner A, da Silva L, Altenbach H (eds) Characterization and development of biosystems and biomaterials. Advanced structured materials. Springer, Berlin, pp 33–54. https://doi.org/10.1007/978-3-642-31470-4_3
124. Charbe N, McCarron PA, Tambuwala MM (2017) Three-dimensional bio-printing: a new frontier in oncology research. World J Clin Oncol 8(1):21–36
125. Zhao Y, Yao R, Ouyang L et al (2014) Three-dimensional printing of Hela cells for cervical tumor model in vitro. Biofabrication 6(3):035001
126. Xu F, Celli J, Rizvi I, Moon S, Hasan T, Demirci U (2011) A three-dimensional in vitro ovarian cancer coculture model using a high-throughput cell patterning platform. Biotechnol J 6(2):204–212
127. Kenny HA, Krausz T, Yamada SD, Lengyel E (2007) Use of a novel 3D culture model to elucidate the role of mesothelial cells, fibroblasts and extra-cellular matrices on adhesion and invasion of ovarian cancer cells to the omentum. Int J Cancer 121:1463–1472. [PubMed: 17546601]
128. Serrano D, Terres MC, Lalatsa A (2018) Applications of 3D printing in cancer. J 3D Print Med 2(3):115–127
129. Albritton JL, Miller JS (2017) 3D bioprinting: improving in vitro models of metastasis with heterogeneous tumor microenvironments. Dis Model Mech 10(1):3–14
130. Zhou X, Zhu W, Nowicki M, Miao S, Cui H, Holmes B, Glazer RI, Grace Zhang LG (2016) 3D bioprinting a cell-laden bone matrix for breast cancer metastasis study. ACS Appl Mater Interfaces 8(44):30017–30026. https://doi.org/10.1021/acsami.6b10673
131. Bray LJ, Binner M, Holzheu A, Friedrichs J, Freudenberg U, Hutmacher DW, Werner C (2015) Multi-parametric hydrogels support 3D in vitro bioengineered microenvironment models of tumour angiogenesis. Biomaterials 53:609–620. https://doi.org/10.1016/j.biomaterials.2015.02.124
132. Pathi SP, Kowalczewski C, Tadipatri R, Fischbach CA (2010) A novel 3-D mineralized tumor model to study breast cancer bone metastasis. PLoS One 5(1):e8849
133. Arrigoni C, Bersini S, Gilardi M, Moretti M (2016) In vitro coculture models of breast cancer metastatic progression towards bone. Int J Mol Sci 17(9):1405
134. Mironov V, Visconti RP, Kasyanov V, Forgacs G et al (2009) Organ printing: tissue spheroids as building blocks. Biomaterials:30, 2164–2174. [PubMed: 19176247]
135. Young EW (2013) Cells, tissues, and organs on chips: challenges and opportunities for the cancer tumor microenvironment. Integr Biol 5:1096–1109
136. Vidi PA, Maleki T, Ochoa M et al (2014) Disease-on-a-chip: mimicry of tumor growth in mammary ducts. Lab Chip 14(1):172–177
137. Guillemot F et al (2010) High-throughput laser printing of cells and biomaterials for tissue engineering. Acta Biomater 6:2494–2500

Chapter 7
Three-Dimensional Bioprinting: Safety, Ethical, and Regulatory Considerations

Ippokratis Pountos, Nazzar Tellisi, and Nureddin Ashammakhi

Abstract Three-dimensional (3D) bioprinting of tissues or organs holds great potential for several clinical applications in the future. Similar to all new biotechnologies, 3D bioprinting possesses both benefits and risks. Consequently, several ethical, safety, and regulatory issues have to be addressed. Ethical concerns identified involve the ownership of prototypes, harvesting and type of cells and biomaterials, research as well as commercialization of produced constructs. Safety concerns identified are linked to the biocompatibility of bioinks, ex vivo manipulation of cells, and maintenance of aseptic conditions. Regulations are vague and are under the provisions made for tissue engineering. Three-dimensional bioprinting should be considered beyond a conceptual therapy; it would require ethical oversight and the introduction of a robust regulatory framework.

Keywords 3D bioprinting · Ethics · Safety · Tissue regeneration · Organs

---

I. Pountos (✉) · N. Tellisi
Academic Department of Trauma and Orthopaedics, Leeds Teaching Hospitals, University of Leeds, Leeds, UK

Chapel Allerton Hospital, Leeds Teaching Hospitals, Leeds, UK
e-mail: pountos@doctors.org.uk

N. Ashammakhi
Center for Minimally Invasive Therapeutics (C-MIT), University of California Los Angeles, Los Angeles, CA, USA

California NanoSystems Institute (CNSI), University of California Los Angeles, Los Angeles, CA, USA

Department of Radiological Sciences, David Geffen School of Medicine, University of California Los Angeles, Los Angeles, CA, USA

Biotechnology Research Center, Libyan Authority for Research, Science and Technology, Tripoli, Libya
e-mail: n.ashammakhi@ucla.edu

M. Guvendiren (ed.), *3D Bioprinting in Medicine*,
https://doi.org/10.1007/978-3-030-23906-0_7

7.1 Introduction

Three-dimensional (3D) printing was originally developed as an industrial tool to facilitate the development of concept-to-prototype designs. From the initial development in aerospace and automotive industries over 40 years ago [1], 3D printing was soon adopted in the medical field for prototyping medical devices and wearable protheses, fabrication of 3D models of human anatomy and instruments to facilitate surgical procedures, training of surgeons [2, 3], and the production of custom-made implants [2]. In parallel, 3D bioprinting involves the use of cell-containing biomaterials to produce constructs that can imitate natural human tissues, and is evolving rapidly [4–7]. There have been multiple clinical applications so far to show the potential of this technology [4, 6]. So, 3D bioprinting will be a major source for repairing tissues and organs and for developing novel regenerative therapeutics [8].

Bioprinting of tissues can be broadly divided into three types, biomimicry, autonomous self-assembly, and the formation of mini-tissues [4, 5]. Biomimicry is the printing of a tissue identical to the host tissue. This is a like-to-like printing where cells, growth factors, and the whole extracellular matrix are printed. Autonomous self-assembly uses cells to create the extracellular matrix and gradually form the whole structure or organ. Mini-tissue formation involves the printing of simple structures such as heart valves, ligaments, or tendons, which will encourage cell invasion and growth. Despite the process used for printing human tissues, several ethical and safety concerns are apparent. This chapter aims to discuss some of the key ethical and safety challenges that 3D bioprinting poses and some of the current directive and advice from regulatory bodies worldwide.

7.2 Ethical Considerations

Bioprinting is a unique technology, which once it matures, has the potential to allow the fabrication of biomimetic 3D tissues and organs identical to the native ones, including 3D printing in situ [9, 10]. Although there is considerable research in the field today, there is little discussion of the socio-ethical aspect of this technology. A number of ethical concerns can be identified regarding the ownership of prototypes, harvest and type of cells and biomaterials, research and commercialization of grafts.

7.2.1 Ownership of Prototypes

It can be claimed that bioprinting enters a new virgin territory regarding legal and ethical considerations. One of the biggest challenges lies with the ownership of the prototypes. There are already concerns raised from law firms that "Inventors of novel bioprinted materials and devices likely have significant concerns about piracy, quality control, and unauthorized products, so it will be critical for them to actively pursue patent protection" [6]. But whose intellectual property is the way our body is built? Is a human organ or a tissue a patentable entity or does it fall to the non-patentable medical techniques category (under the medical treatment exemption)? In the publication of JL Tran it is stated that bioprinting technique and products are two different entities [7]. Although a bioprinting process can be patentable, with some patents already approved, bioprinting human products is largely controversial. If such practice is allowed, monopoly over grafts and organs can be the ultimate outcome, which will lead to social stratification and unfair use of this technology. On the other hand, who can bear the increased cost for the research and advancement of this technology, other than companies aiming for future profits? The International Federation of Associations of Anatomy states there should be no commercialization of human parts for education and research [11]. So, can human structure prototypes be sold for profit? If, for instance, it is decided that human organ structures are not patentable, what can happen for alterations? Will it be allowed and if allowed surely there is a case of patentable entity (for example, skin with enhanced dermis or epidermis or hearts with improved valves). It should be first decided how the prototypes of tissues/organs will be acquired and scanned. Human anatomy is not a one-fit-all construct. Variations regarding the structure of the human body between people exist and even having a good idea of the histological structure of such tissues, little is known of the myriad interactions of cells with their environment. So, considerable attention should be paid to advancing our understanding of the basic biology.

Another factor to consider is that even if we produce 3D printed tissues and organs from human cells, who can claim ownership of the final product? Is it the donor of the cells, the clinician, the company, or the institution? [12–14] Also, there are several safe-keeping issues that should not be underestimated. Predictions estimate that approximately $100 billion are lost per year in intellectual property [6]. This raises concerns in terms of who will use this technology and how. The Liberator, the first 3D-printable gun whose CAD files were released over the internet, is a reality. So, there is a real danger that bioprinting of prototypes of organs and complex structures can be weaponized by bioterrorism.

7.2.2 *Cells*

Bioprinting live tissues prerequisites incorporation of living cells. Most likely, patient's own cells won't be available or if available, would have suboptimal properties and be of poor regeneration capacity. An example is the bone marrow-derived stem cells from elderly patients, which is shown to have significant deficiencies in function compared to those of younger individuals. In addition, the level of cell differentiation required for the printed cells is unknown. Already differentiated cells might be of poor function and adult stem cells will have to be included in the prints. But, adult stem cells are rare and their ex vivo manipulation can make them lose some of their properties [15]. Hence, utilizing embryonic stem cells might raise significant ethical and safety issues. Removing these stem cells requires the destruction of the embryo. Such practice has already triggered significant controversy as it can be considered as destruction of a human being. Although many countries issued a legal framework on this subject, deviations exist. In 2000 for example, the UK Government voted to allow the use of somatic cell nuclear replacement [16]. In this rather controversial vote, genetically matched patient-specific stem cells could have been produced from cloned embryos. Besides the ethical issues, this program was criticized to be costly and impractical [16].

Another potential area of concern is the use of fetal cells and tissues from elective abortions (also ectopic pregnancies, spontaneous abortions, or even stillbirths). Fetal-tissue implants appeared to offer some clinical benefit to some conditions, such as advanced Parkinson's disease [17, 18]. Some claim this is the first step towards the commercial use of such tissues [17]. Scepticists suggest that it is morally abhorrent for conception to take place to produce a fetus for abortion and subsequent use or sale for cells, organ or tissue donations [17].

Regardless of the source of cells, questions arise about the ownership and overall costs involved. Usually, worldwide it is illegal to buy or sell human organs, tissues, and cells. However, "reasonable fees" can be charged for the processing of these cells. So financial benefits exist. In addition, incidents involving non-consented procurement, inadequate testing, inaccurate or false donor files, irresponsible allocations, and illegal trafficking of human cells, tissues, and products have been reported [19, 20]. As with human organs and tissues, is it ethical or even legal to patent stem cell lines? Definitely, it is not in the public interest to allow such practice [21]. The second problem lies with the ownership of these cells; is it the clinician, the healthcare institution, or the bioprinting company that can claim ownership of the cells? This is an ongoing issue faced in tissue engineering, which will require further clarification and guidance when more complex tissues are fabricated. And finally, what shall the consent process involve? Shall the clinician or surgeon be involved or separated from the research process?

7.2.3 Justice

In today's organ transplantation programs, provisions for fairly distributing organs exist. These provisions consider the lifesaving nature of the transplantation, and the probability of a good outcome is highly emphasized to achieve the maximum benefit for all transplantations. Having the ability to print organs or tissues on demand with their potential commercialization can restrict the use of this technology to those unable to afford it. So, from organ transplantation being of an altruistic nature, it could become a commodity for those who can pay for their organs. Perhaps those who can pay for such services will live longer, have a better quality of life and avoid the consequences of poor health or the need of prolonged drug therapies (for example, insulin replacement). It could also mean that the financially unfortunate patients must still wait for their "secondhand" organs or get exposed to complex operations (for instance, skin flaps or autologous skin grafts for skin loss). Even with a tight regulatory framework, black markets of questionable-quality bioprinted organs or tissues might thrive. Such unregulated printing of tissues could find applications in many controversial scenarios like cosmetic surgeries or performance enhancement in elite athletes [6, 22].

7.2.4 Degradation of Social, Moral, and Religious Beliefs and Values

Perhaps on-demand bioprinting of organs and tissues could cause a change in our views for life. If we see ourselves as the modulators of life, can we then lose reverence for life? [23] Moral degradation can occur when the availability of human parts changes our perception of unhealthy habits and practices (i.e., smoking, alcohol, etc.). It could be claimed that we will transform the mystery and majesty of life into a mere malleable and potentially marketable commodity. This argument could apply to Louise Brown, the first baby born using in vitro fertilization (IVF). IVF treatment is widely accepted today, but then, some people considered it a breakthrough achievement while others considered it as the creation of a monster.

Ethical concerns related to the clinical translations also exist. Today's scientists and academics experience high pressure from the industry and universities for the production of breakthrough discoveries and high impact research [24]. Exaggerated claims and loss of their independence is often the case. If, for example, we take the clinical use of mesenchymal stem cells (MSCs), would someone expect to analyze their properties first before they are introduced to the clinical practice? To date, the majority of research on MSCs has been carried out on animal cells, which have different characteristics to that of humans. We have not even identified a marker which defines MSCs, hence International Society of Cell Therapies (ISCT) has issued broad guidelines on the minimal criteria and properties that MSC should pose (adherence to tissue culture plastic, phenotype and differentiation potential) [25]. With such little knowledge about these cells, MSCs are used, sold, and implanted clinically and commercially. This raises significant ethical questions for those involved in such programs and potential safety concerns for the patients. On the other hand, if industry is not involved, who is going to financially support the bioprinting technology?

Concerns derived from religious beliefs also exist. Some religions already prohibit therapeutic and reproductive cell cloning (Catholic, Orthodox, and Islam). Some prohibit therapeutic cloning (Buddhists) while others accept them (Jewish). It is yet unknown how religious groups worldwide will perceive bioprinting.

7.3 Safety

7.3.1 Safety: Ex Vivo Manipulation of Human Cells

The homeostasis and fate of the cells are strongly regulated by their environment. Outside the body, the properties of the artificial cellular environment are crucial. This environment should make handling the cells easy and should allow the cells to survive and even proliferate, generating high yields. It is then crucial to optimize, maintain, and monitor the culture conditions to expand the cells. One of the major challenges is for the system to be neutral and to not alter the phenotype of the cells and not induce karyotypic or genotypic abnormalities.

7.3.1.1 Malignant Transformation

Ex vivo culture and manipulation can cause early cellular senescence [26–33]. However, there is evidence that human stem cells could transform to malignant cells in culture following chromosomal alterations and re-arrangements [26–31]. Rosland et al. have shown rates of spontaneous malignant transformation in bone marrow-derived MSCs in 11 out of 24 cultures (45.8%) [26]. High frequency of karyotype alterations were also seen in cord blood endothelial progenitor cells [31]. Similar results were reported by other authors but the significance of these results requires further elucidation [27–30].

7.3.1.2 Culture Conditions

Besides the genotypic alterations, the culture conditions are crucial and can alter the cellular behavior and fate. The work of Sotiropoulou et al. showed that the quality and/or the properties of standard commercially available tissue culture plastics used to expand MSCs is crucial for their proliferation ex vivo [32]. In addition, the same cells in culture seem to alter their phenotype through passaging and also lose their ability to differentiate towards specific phenotypes (cartilage and bone) while they preserve their ability to become adipocytes [33, 34]. Schallmoser et al. have shown significant gene expression changes with cell culture [34]. In particular, cell culture was found to upregulate apoptosis, differentiation, and senescence while the genes involved in mitosis and proliferation were downregulated [34]. Other studies have shown that the phenotype of stem cells changes in cell culture even in early passages [29]. Therefore, besides the risk for potential tumorigenic phenotypic changes, it is then questionable whether, in a potential future application, the cells implanted would be the same or as effective as the native cells of the human body.

7.3.1.3 Culture Media

Culture media are essential for cell manipulation outside the human body. These media stabilize the cells in culture allowing their survival and attachment, and deliver macromolecules, trophic factors, minerals, proteins, and trace elements [35–37]. These media can cause toxicity and also inhibit specific functions of the cells [33]. More specificaly they can for example, inhibit differentiation and proteases production; and disturb the maintainance of ideal pH levels; inhibit proteases; and are toxic for the cells' molecules [33]. Fetal animal sera have been traditionally used as it contains low levels of gamma (γ)-globulins, hence it has a reduced risk for antibody interactions with the cells [16].

The ideal composition of culture media in tissue engineering and stem cell research has been a subject of vivid discussions over the last decades [33]. The cells used in 3D bioprinting would have to be handled, manipulated, and stored ex vivo. Ex vivo expansion prerequisites the use of media first for handling the cells and secondly to support the growth of the stem cells in culture including media that contain essential cues for cell growth [28]. For many cell types and especially stem cells, culture media have to contain animal fetal sera with a concentration of 10–20% [33]. Although these sera are heat inactivated, there is still a risk of transmission of prions, zoonoses and immune reactions [33]. Research has shown that the fetal calf serum used in clinical applications of cardiomyoplasty with the use of MSCs resulted in episodes of life-threatening arrhythmias [38]. In addition, other studies have reported immune reactions to fetal calf proteins [39, 40]. The risk for transmissible spongiform encephalopathies was also reported [41]. Apart from the risk of immunological reactions, the lack of standardization by the companies has led to great variability in their performance [42]. Seasonal and continental variations and batch-to-batch discrepancies were also reported [42]. More recently, knockout serum replacement replaced fetal bovine sera as an equivalent but safer alternative [43]. These sera, however, still contain animal-derived components. Other alternatives include the use of media such as the mTesR1 (Stem Cell Technologies) and StemPro (ThermoFisher) which are serum-free. However, these media contain bovine serum albumin [44]. At present there is extensive research on xeno-free media, but such products will require extensive validation [45, 46].

Using human blood products as substitutes for animal sera have been also explored [33, 47–50]. The use of human serum to support the culture of human cells in vitro is affected by the state of the body at the time of collection [51, 52]. Our research on the effect of autologous serum derived from patients after bone fractures has shown great variability on the proliferative and osteogenic capacity of the cells when exposed to serums taken at different time points following the injury [51, 53]. In this study, significant changes in the growth factor content were noted even within a 24 h period which affected the cell growth kinetics and differentiation [51]. Similar observations were reported in relation to donor age, with blood from younger donors outperformed that of older in age [52]. Hence, it is understandable that literature is controversial with a number of studies proposing that human blood products are inferior to fetal animal sera [33, 49–51]. An alternative approach is the use of human platelet rich plasma (PRP). Using PRP has shown satisfactory results [47, 51]. However, drawbacks similar to those associated with the use of human serum do exist, adding more obstacles to the limited availability of platelet units [47, 52].

7.3.1.4 Sterility

Effective strategies for pathogen elimination should be implemented in all the steps of bioprinting starting from the isolation of cells and materials to the ex vivo manipulation of cells and the actual bioprinting of human tissues. In such a complex process, reducing the viral and bacterial contamination is a challenge. Maintaining sterility and sterilization of 3D printed implants is a vital factor when it comes to clinical applications. Today's technologies include solvent detergent treatment, methylene blue with light, amotosalen or psoralen or rivofloxacin with ultraviolet light [54, 55]. It is unknot known what the effect of these products is on the human body and cells exposed to these treatments. Evidence shows that such treatments could inhibit cellular growth ex vivo and are linked to genotoxicity, mutagenicity, neoantigenicity, reproductive toxicity, and fetal developmental problems [56–60].

7.3.2 Safety: Use of Growth Factors

Three-dimensional bioprinting of human tissues could potentially include signals to the cells from cytokines and growth factors. Although many cytokines and growth factors are commercially available, it is not known what their effect on the cells and a human body is, in general. There is evidence to support the view that exposure to some growth factors can irreversibly alter the faith of the cells. For example, fibroblast growth factor-2 was recently found to induce malignant transformation of MSCs [61]. Similarly, clinical studies have shown that high doses of bone morphogenetic proteins (BMPs) used in the treatment of fractures, bony fusions, and nonunions can increase the risk for malignancies [62]. Our knowledge regarding the effect of cytokines and growth factors is limited. It is unknown and difficult to quantify the ideal concentrations of growth factors required at a cellular level. Taking BMP-2 as an example, a molecule commercialized to miraculously induce bone growth about 15 years ago, the results today are mixed [2, 63]. There are reports of significant complications, mixed clinical results equivalent to pre-existing low-cost techniques, and lack of understanding of its long-term effects [63]. This is despite numerous in vitro and pre-clinical animal studies that failed to capture these events and uniformly praised its use in bone healing.

7.4 Regulation

3D printed human tissues are a reality that gradually evolves. Companies such as the Organovo Inc. claim they managed to bioprint human liver and kidneys (ExVive™ human liver or kidney tissue) that are offered for contract testing services. However, the regulatory framework under which these tissues and services are evolving is vague and partially captured by the tissue engineering regulations [64]. Exaggerated claims and misconducts are a reality. A recent example is the case of potential fraud involving FBS label non-conformances in the USA during the period of 2003–2011 [65, 66]. During this period, approximately 280,000 L of FBS which contained added adult bovine serum albumin, water, and cell growth promoting additives were sold [65, 66]. When they were discovered, the FDA recalled these products [65, 66].

There are several challenges which are faced, with an attempt to regulate bioprinting. Take the FDA as an example of who is responsible for controlling and distributing medical products in the USA (Fig. 7.1) [67]. At first, issues will arise with the classification of these products as this dictates the pathway for the review and approval of such devices. A complex bioprinting product is neither a device, biologic or drug. Perhaps it may or may not fall under the generic and vague category "combination product." So, do we need a new category to capture such products? Adding to the complexities, the FDA vaguely mentions that some products "might necessitate additional manufacturing process considerations and different regulatory pathways". However, often the FDA involves the CBER (Centre for Biologics Evaluation and Research) in the decision process but at present, there are no specific provisions for 3D printed organs or grafts. In Australia, the Therapeutic

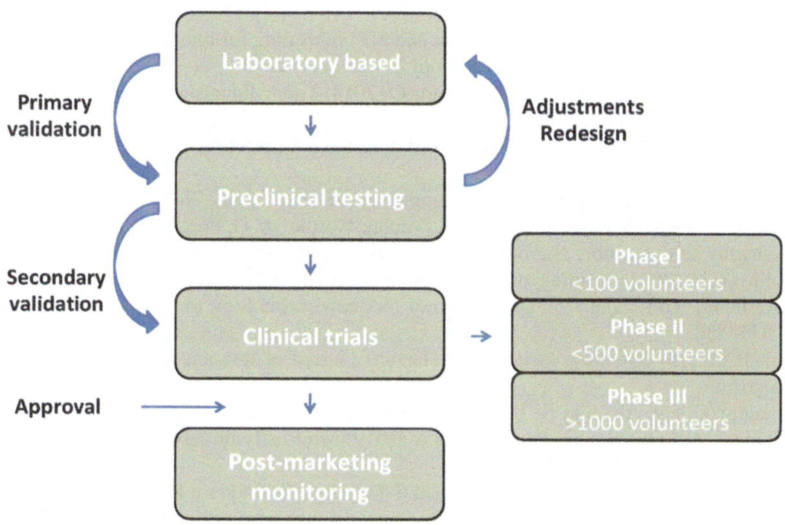

Fig. 7.1 FDA review process

Goods Administration allows custom-made devices to be made without premarket assessment. That means that if a 3D printed graft or organ falls within the device category, it can then be fabricated with no control. There are similar problems in Europe where bioprinting can be regulated under the provisions made for tissue engineering [68]. In Asia, religious norms additionally influence the approval process [69]. It is anticipated that the authorities and regulatory bodies across the globe must discuss and conclude on the potential risks and conclude on the regulatory framework governing this emerging technology [70].

7.5 Conclusions

3D bioprinting technology has the potential to produce complex tissue constructs in the near future. However, several ethical and safety concerns are apparent. A strict ethical framework regarding their commercialization should be put in place. Basic research on the ex vivo cell handling and manipulation is required to assure safety of the final products. Finally, the appropriate regulatory bodies worldwide should produce advice and directives, as this technology is not currently captured in the current legislation and regulatory frameworks at present.

Conflict of Interest No benefits in any form have been received or will be received from a commercial party related directly or indirectly to the subject of this chapter.

References

1. Ghidini T (2018) Regenerative medicine and 3D bioprinting for human space exploration and planet colonisation. J Thorac Dis 10(Suppl 20):S2363–S2375
2. Tellisi N, Ashammakhi NA, Billi F, Kaarela O (2018) Three dimensional printed bone implants in the clinic. J Craniofac Surg 29:2363–2367
3. Ashammakhi N, Kaarela O (2018) Three-dimensional bioprinting can help bone. J Craniofac Surg 29:9–11
4. Murphy SV, Atala A (2014) 3D bioprinting of tissues and organs. Nat Biotechnol 32:773–785
5. Jakab K, Norotte C, Marga F, Murphy K, Vunjak-Novakovic G, Forgacs G (2010) Tissue engineering by self-assembly and bio-printing of living cells. Biofabrication 2:022001
6. Wolinsky H (2014) Printing organs cell-by-cell: 3-D printing is growing in popularity, but how should we regulate the application of this new technology to health care? EMBO Rep 15:836–838
7. Tran JL (2015) Patenting bioprinting. Harvard journal of law and technology digest, 2015 symposium
8. Ashammakhi N, Ahadian S, Darabi MA, El Tahchi M, Lee J, Suthiwanich K, Sheikhi A, Dokmeci MR, Oklu R, Khademhosseini A (2019) Minimally invasive and regenerative therapeutics. Adv Mater 31(1):e1804041
9. Ashammakhi N, Ahadian S, Pountos I, Hu S-K, Tellisi N, Bandaru P, Ostrovidov S, Dokmeci M, Khademhosseini A (2019) In situ three-dimensional printing for reparative and regenerative therapy. Biomed Microdevices 21(2):42

10. Ashammakhi N, Kaarela O, Hasan A, Byambaa B, Sheikhi A, Gaharwar AK, Khademhosseini A (2019) Advancing frontiers in bone bioprinting. Adv Healthc Mater 8(7):e1801048
11. Recommendations of the IFAA. http://www.ifaa.net/recommendations/
12. Awaya T (2005) Common ethical issues in regenerative medicine. J Int Bioethique 16:69–75
13. Black J (1997) Thinking twice about "tissue engineering" [Ethical issues]. Eng Med Biol Mag IEEE 16:102–104
14. Samanta A, Samanta JO, Price D (2004) Who owns my body—thee or me? The human tissue story continues. Clin Med 4:327–331
15. Prockop DJ, Olson SD (2007) Clinical trials with adult stem/progenitor cells for tissue repair: let's not overlook some essential precautions. Blood 109:3147–3151
16. Wilmut I (2004) The moral imperative for human cloning. New Sci 181:16–17
17. Spencer DD, Robbins RJ, Naftolin F, Marek KL, Vollmer T, Leranth C, Roth RH, Price LH, Gjedde A, Bunney BS (1992) Unilateral transplantation of human fetal mesencephalic tissue into the caudate nucleus of patients with Parkinson's disease. N Engl J Med 327:1541–1548
18. Freed CR, Breeze RE, Rosenberg NL, Schneck SA, Kriek E, Qi JX, Lone T, Zhang YB, Snyder JA, Wells TH (1992) Survival of implanted fetal dopamine cells and neurologic improvement 12 to 46 months after transplantation for Parkinson's disease. N Engl J Med 327:1549–1555
19. Collins S (2001) Tissue banks: is the Federal Government's oversight adequate? Hearing before the Committee on Governmental Affairs, US Senate, vol 264. Diane Publishing Company, Washington, DC
20. Pirnay JP, Vanderkelen A, Zizi M, De Vos D, Rose T, Laire G, Ectors N, Verbeken G (2010) Human cells and tissues: the need for a global ethical framework. Bull World Health Organ 88:870–872
21. European Group on Ethics in Science and New Technologies (2002) Opinion 16: ethical aspects of patenting inventions involving human stem cells, 7 May 2002, European Commission, Brussels
22. Ashammakhi N, Darabi MA, Pountos I (2019) The dynamic cycle of future personalized and regenerative therapy. J Craniofac Surg 30(3):623–625
23. Boenink M, Swierstra T, Stemerding D (2010) Anticipating the interaction between technology and morality: a scenario study of experimenting with humans in bionanotechnology. Stud Ethics Law Technol 4:1–38
24. House of Lords (2000) Science and Society, report of the House of Lords Select Committee on Science and Technology. HMSO, London
25. Dominici M, Le Blanc K, Mueller I, Slaper-Cortenbach I, Marini F, Krause D, Deans R, Keating A, Prockop DJ, Horwitz E (2006) Minimal criteria for defining multipotent mesenchymal stromal cells. The International Society for Cellular Therapy position statement. Cytotherapy 8:315–317
26. Rosland GV, Svendsen A, Torsvik A, Sobala E, McCormack E (2009) Long-term cultures of bone marrow-derived human mesenchymal stem cells frequently undergo spontaneous malignant transformation. Cancer Res 69:5331–5339
27. Miura M, Miura Y, Padilla Nash HM, Molinolo AA, Fu B (2006) Accumulated chromosomal instability in murine bone marrow mesenchymal stem cells leads to malignant transformation. Stem Cells 24:1095–1103
28. Li H, Fan X, Kovi RC, Jo Y, Moquin B (2007) Spontaneous expression of embryonic factors and p53 point mutations in aged mesenchymal stem cells: a model of age-related tumorigenesis in mice. Cancer Res 67:10889–10898
29. Popov BV, Petrov NS, Mikhaĭlov VM, Tomilin AN, Alekseenko LL, Grinchuk TM, Zaĭchik AM (2009) Spontaneous transformation and immortalization of mesenchymal stem cells in vitro. Tsitologiia 51:91–102
30. Rubio D, Garcia-Castro J, Martin MC, Fuente R, Cigudosa JC (2005) Spontaneous human adult stem cell transformation. Cancer Res 65:3035–3094
31. Corselli M, Parodi A, Mogni M, Sessarego N, Kunkl A, Dagna-Bricarelli F, Ibatici A, Pozzi S, Bacigalupo A, Frassoni F, Piaggio G (2008) Clinical scale ex vivo expansion of cord blood-

derived outgrowth endothelial progenitor cells is associated with high incidence of karyotype aberrations. Exp Hematol 36:340–349
32. Sotiropoulou PA, Perez SA, Salagianni M, Baxevanis CN, Papamichail M (2006) Characterization of the optimal culture conditions for clinical scale production of human mesenchymal stem cells. Stem Cells 24:462–471
33. Pountos I, Corscadden D, Emery P, Giannoudis PV (2007) Mesenchymal stem cell tissue engineering: techniques for isolation, expansion and application. Injury 38(Suppl 4):S23–S33
34. Schallmoser K, Bartmann C, Rohde E, Bork S, Guelly C, Obenauf AC, Reinisch A, Horn P, Ho AD, Strunk D, Wagner W (2010) Replicative senescence-associated gene expression changes in mesenchymal stromal cells are similar under different culture conditions. Haematologica 95:867–874
35. van der Valk J, Gstraunthaler G (2017) Fetal bovine serum (FBS)—a pain in the dish? Altern Lab Anim 45:329–332
36. Gstraunthaler G (2003) Alternatives to the use of fetal bovine serum: serum-free cell culture. ALTEX 20:275–281
37. van der Valk J, Brunner D, De Smet K, Fex Svenningsen A, Honegger P, Knudsen LE, Lindl T, Noraberg J, Price A, Scarino ML, Gstraunthaler G (2010) Optimization of chemically defined cell culture media—replacing fetal bovine serum in mammalian in vitro methods. Toxicol In Vitro 24:1053–1063
38. Chachques JC, Herreros J, Trainini J, Juffe A, Rendal E, Prosper F, Genovese J (2004) Autologous human serum for cell culture avoids the implantation of cardioverter-defibrillators in cellular cardiomyoplasty. Int J Cardiol 95(Suppl 1):S29–S33
39. Selvaggi TA, Walker RE, Fleisher TA (1997) Development of antibodies to fetal calf serum with arthus-like reactions in human immunodeficiency virus-infected patients given syngeneic lymphocyte infusions. Blood 89:776–779
40. Tuschong L, Soenen SL, Blaese RM, Candotti F, Muul LM (2002) Immune response to fetal calf serum by two adenosine deaminase-deficient patients after T cell gene therapy. Hum Gene Ther 13:1605–1610
41. WHO. Guidelines on tissue infectivity distribution in transmissible spongiform encephalopathies. http://www.who.int/bloodproducts/cs/TSEPUBLISHEDREPORT.pdf?ua=1
42. Dimasi L (2011) Meeting increased demands on cell-based processes by using defined media supplements. Bioprocess J 9:8
43. Xu C, Rosler E, Jiang J, Lebkowski JS, Gold JD, O'Sullivan C, Delavan-Boorsma K, Mok M, Bronstein A, Carpenter MK (2005) Basic fibroblast growth factor supports undifferentiated human embryonic stem cell growth without conditioned medium. Stem Cells 23:315–323
44. Adil M, Schaffer DV (2017) Expansion of human pluripotent stem cells. Curr Opin Chem Eng 15:24–35
45. Karnieli O, Friedner OM, Allickson JG, Zhang N, Jung S, Fiorentini D, Abraham E, Eaker SS, Yong TK, Chan A, Griffiths S, Wehn AK, Oh S, Karnieli O (2017) A consensus introduction to serum replacements and serum-free media for cellular therapies. Cytotherapy 19(2):155–169
46. Cimino M, Gonçalves RM, Barrias CC, Martins MCL (2017) Xeno-free strategies for safe human mesenchymal stem/stromal cell expansion: supplements and coatings. Stem Cells Int 2017:6597815
47. Burnouf T, Strunk D, Koh MB, Schallmoser K (2016) Human platelet lysate: replacing fetal bovine serum as a gold standard for human cell propagation? Biomaterials 76:371–387
48. Koller MR, Maher RJ, Manchel I, Oxender M, Smith AK (1998) Alternatives to animal sera for human bone marrow cell expansion: human serum and serum-free media. J Hematother 7:413–423
49. Lin HT, Tarng YW, Chen YC, Kao CL, Hsu CJ, Shyr YM, Ku HH, Chiou SH (2005) Using human plasma supplemented medium to cultivate human bone marrow-derived mesenchymal stem cell and evaluation of its multiple-lineage potential. Transplant Proc 37:4504–4505

50. Stute N, Holtz K, Bubenheim M, Lange C, Blake F, Zander AR (2004) Autologous serum for isolation and expansion of human mesenchymal stem cells for clinical use. Exp Hematol 32:1212–1225
51. Lohmann M, Walenda G, Hemeda H, Joussen S, Drescher W, Jockenhoevel S, Hutschenreuter G, Zenke M, Wagner W (2012) Donor age of human platelet lysate affects proliferation and differentiation of mesenchymal stem cells. PLoS One 7(5):e37839
52. Pountos I, Georgouli T, Henshaw K, Bird H, Giannoudis PV (2013) Release of growth factors and the effect of age, sex, and severity of injury after long bone fracture. A preliminary report. Acta Orthop 84:65–70
53. Pountos I, Georgouli T, Giannoudis PV (2008) The effect of autologous serum obtained after fracture on the proliferation and osteogenic differentiation of mesenchymal stem cells. Cell Mol Biol (Noisy-le-Grand) 54:33–39
54. Astori G, Amati E, Bambi F, Bernardi M, Chieregato K, Schäfer R, Sella S, Rodeghiero F (2016) Platelet lysate as a substitute for animal serum for the ex-vivo expansion of mesenchymal stem/stromal cells: present and future. Stem Cell Res Ther 7:93
55. Salunkhe V, van der Meer PF, de Korte D, Seghatchian J, Gutiérrez L (2015) Development of blood transfusion product pathogen reduction treatments: a review of methods, current applications and demands. Transfus Apher Sci 52:19–34
56. Mundt JM, Rouse L, Van den Bossche J, Goodrich RP (2014) Chemical and biological mechanisms of pathogen reduction technologies. Photochem Photobiol 90:957–964
57. Goodrich RP, Segatchian J (2018) Special considerations for the use of pathogen reduced blood components in pediatric patients: an overview. Transfus Apher Sci 57:374–377
58. Reddy HL, Dayan AD, Cavagnaro J, Gad S, Li J, Goodrich RP (2008) Toxicity testing of a novel riboflavin-based technology for pathogen reduction and white blood cell inactivation. Transfus Med Rev 22:133–153
59. Ciaravi V, McCullough T, Dayan AD (2001) Pharmacokinetic and toxicology assessment of INTERCEPT (S-59 and UVA treated) platelets. Hum Exp Toxicol 20:533–550
60. Pountos I, Georgouli T, Henshaw K, Howard B, Giannoudis PV (2014) Mesenchymal stem cell physiology can be affected by antibiotics: an in vitro study. Cell Mol Biol (Noisy-le-Grand) 60:1–7
61. Yamaoka E, Hiyama E, Sotomaru Y, Onitake Y, Fukuba I, Sudo T, Sueda T, Hiyama K (2011) Neoplastic transformation by TERT in FGF-2-expanded human mesenchymal stem cells. Int J Oncol 39:5–11
62. Pountos I, Panteli M, Georgouli T, Giannoudis PV (2014) Neoplasia following use of BMPs: is there an increased risk? Expert Opin Drug Saf 13:1525–1534
63. Barcak EA, Beebe MJ (2017) Bone morphogenetic protein: is there still a role in orthopedic trauma in 2017? Orthop Clin North Am 48:301–309
64. Gilbert F, O'Connell CD, Mladenovska T, Dodds S (2018) Print me an organ? Ethical and regulatory issues emerging from 3D bioprinting in medicine. Sci Eng Ethics 24:73–91
65. Gstraunthaler G, Lindl T, van der Valk J (2013) A plea to reduce or replace fetal bovine serum in cell culture media. Cytotechnology 65:791–793
66. Gstraunthaler G, Lindl T, van der Valk J (2014) A severe case of fraudulent blending of fetal bovine serum strengthens the case for serum-free cell and tissue culture applications. Altern Lab Anim 42:207–209
67. Smith DS (2006) The Government's role in advancing regenerative medicine and tissue engineering–science, safety, and ethics. Periodontology 41:16–29
68. Heinonen M, Oila O, Nordström K (2005) Current issues in the regulation of human tissue-engineering products in the European Union. Tissue Eng 11:1905–1911
69. Gheisari Y, Baharvand H, NayerniaK VM (2012) Stem cell and tissue engineering research in the Islamic republic of Iran. Stem Cell Rev Rep 8:629–639
70. Neely EL (2016) The risks of revolution: ethical dilemmas in 3D printing from a US perspective. Sci Eng Ethics 22:1285–1297

Correction to: 3D Bioprinting Technologies

Christopher B. Highley

Correction to: Chapter 1 in: M. Guvendiren (ed.), *3D Bioprinting in Medicine*, https://doi.org/10.1007/978-3-030-23906-0_1

This book was inadvertently published without updating the corrections to chapter 1. The same has been updated in all renditions of the book.

The updated online version of this chapter can be found at
https://doi.org/10.1007/978-3-030-23906-0_1

M. Guvendiren (ed.), *3D Bioprinting in Medicine*,
https://doi.org/10.1007/978-3-030-23906-0_8

Index

M. Guvendiren (ed.), *3D Bioprinting in Medicine*,
https://doi.org/10.1007/978-3-030-23906-0

The manufacturer's authorised representative in the EU is Springer Nature Customer Service Centre GmbH, Europaplatz 3, 69115 Heidelberg, Germany. If you have any concerns regarding our products, please contact ProductSafety@springernature.com

Printed and bound by CPI Group (UK) Ltd, Croydon, CR0 4YY
30/09/2026
02225229-0005